Earth and Environmental Sciences Library

Earth and Environmental Sciences Library (EESL) is a multidisciplinary book series focusing on innovative approaches and solid reviews to strengthen the role of the Earth and Environmental Sciences communities, while also providing sound guidance for stakeholders, decision-makers, policymakers, international organizations, and NGOs.

Topics of interest include oceanography, the marine environment, atmospheric sciences, hydrology and soil sciences, geophysics and geology, agriculture, environmental pollution, remote sensing, climate change, water resources, and natural resources management. In pursuit of these topics, the Earth Sciences and Environmental Sciences communities are invited to share their knowledge and expertise in the form of edited books, monographs, and conference proceedings.

More information about this series at http://www.springer.com/series/16637

Kirtikumar Randive · Shubhangi Pingle ·
Anupam Agnihotri
Editors

Innovations in Sustainable Mining

Balancing Environment, Ecology and Economy

Editors
Kirtikumar Randive
Department of Geology
Rashtrasant Tukadoji Maharaj Nagpur University
Nagpur (MH), India

Shubhangi Pingle
Regional Occupational Health Centre (Southern)
National Institute of Occupational Health
Bengaluru (KA), India

Anupam Agnihotri
Jawaharlal Nehru Aluminium Research Development, and Design Centre
Nagpur (MH), India

ISSN 2730-6674 ISSN 2730-6682 (electronic)
Earth and Environmental Sciences Library
ISBN 978-3-030-73798-6 ISBN 978-3-030-73796-2 (eBook)
https://doi.org/10.1007/978-3-030-73796-2

This Springer imprint is published by the registered company Springer Nature Switzerland AG
The registered company address is: Gewerbestrasse 11, 6330 Cham, Switzerland

Foreword

Modern industrial revolution owes a lot to the mining industry since all the raw materials including metals and precious stones are dug out from the mines. However, mining comes with penalties in the form of environmental degradation; reduction in groundwater levels and contamination of air, water and soil. The dilemma which is expressed by the editors in this book is what to choose between mining and environment. The answer is that we need both. Neither we can stop development in the name of environment nor we can ignore environment in the name of development. Therefore, it is imperative that the stakeholders need to find a viable solution in which mining and environment both go hand in hand. This is what we know as sustainability. However, sustainability can be measured in different ways. Whereas the sustainability paradigm draws heavily from the stipulated, time-bound extraction of mineral resources; it also needs to be measured in terms of maximum utilization of ore as well as wastes produced in mines. Excavations for ores generate a huge number of wastes. Managing such wastes itself is a great environmental challenge. Therefore, it is important that the technologies be developed in a manner in which the low-grade ore is converted to high-grade ore, extracting the by-products for increasing profitability, and bringing innovative ideas for creating value-added products. The scientific community is now aware that the natural resources will not last for long; therefore, the concept of zero-waste mining and following 4Rs, namely, reduce, reuse, recycle, and recover are important. This is the need of the hour, and the scientific community and academia are once again asked to play a key role in this new age technological challenge.

The very title of the book 'Innovations in Sustainable mining' attracted me for the reason that both sustainability and innovation go hand in hand. One cannot exist without the other, and this is how humanity progresses. The book contains 14 chapters, each of which is informative and presents the current state of knowledge in the field of sustainable mining and value-added products from the mining waste. Editors of this book are eminent researchers and academicians. Similarly, authors are highly reputed and revered in their respective domains of expertise. The book is being published by Springer which is a leading book publishing company in the world. Therefore, this book achieves high academic standards. I congratulate the authors and editors for bringing out this valuable compilation.

Nagpur, India Dr. Subhash Choudhary

Preface

Mining has always been a critical issue of discussion. On one hand, mining is regarded as the driver of the growth of civilization, whereas, on the other, it is criticized for environmental degradation. It's a difficult choice. The economists, policymakers and entrepreneurs favour continuity of mining activity; however, the environmentalists, social workers and local residents are against it. The major problems associated with mining activities are loss of green cover; contamination and pollution of soil, water and air; depletion of water table; loss of habitat of animals and many socio-economic conflicts associated with mining operations. Nevertheless, there are many benefits of mining, such as direct and indirect employment generation, development of backward areas, construction of civil networks, education, hospitals, markets and so on.

Irrespective of this centuries-long debate, one has to understand that mining is essential for nations to secure needs and safeguard their interests. Therefore, mining activity cannot stop in a foreseeable future. One of the major problems associated with the mining operations is generation of huge overburden, which has to be stored as dumps near mines. Moreover, there are washeries and beneficiation plants in the proximity of mines, which add to environmental woes. The solid waste dumps as well as tailing ponds are the major sources of contamination of the toxic elements that are mined or extracted as by-products, associated products or rejects during beneficiation. Nevertheless, the benefits of mining are overwhelming and cannot be ignored altogether. Thus, it is imperative that a critical balance be maintained between mining and environment. This is why the concept sustainable mining has emerged in the past, and now standing firmly on the crossroads, where the innovations are necessary to make the mining activity sustainable.

Sustainability in the mining can be brought by bringing new innovations in mining technology, new design of mines, reducing the overburden and creating innovative value-added by-products from the mining waste. Secondly, the innovations should also be brought in mitigating the hazards due to mining activity such as degrading environment, depleted water table and contaminated water bodies, remediation of soils in mining areas and so on. With above objectives in mind, the active workers, scientists and academicians were invited to publish their ideas and latest development in the sustainable mining in a compiled volume. This book, therefore, presents new ideas as well as current state of knowledge in the sustainable mining.

The book comprises 14 chapters, first chapter by **Jawadand and Randive** presents a review of the sustainable mining technologies and how the mining waste can be transformed into useful resources. **Suchitra Rai et al.** come out with a novel idea of utilization of waste from aluminium industry in pollution control. Similarly, **Upendra Singh** proposes very innovative products from the aluminium dross, which is an industrial waste. **Najar et al.** also discuss very innovative geopolymer products using industrial waste. More innovation is proposed by **Dandekar et al** using fly ash and bamboo leaves for making the polymeric bricks.

Technological innovations in beneficiation of low-grade ores to high-grade ores greatly help in value addition and sustainability of mines. **Bhukte et al.** proposed new method for beneficiation of low-grade lateritic bauxite ore. **Datta and Nandi** also discussed new technological innovations in beneficiation of bauxite from Guinea. Arsenic is a deleterious environmental contaminant and often causes toxicity by contaminating the groundwater. However, **Bage et al.** have a different look at arsenic in which they used arsenic as a tracer for locating hidden precious metal (gold) deposit. Nowadays, the rare earth elements are most sought-after industrial minerals. However, the deposits of REE are only meagre compared to their demands. Therefore, the innovative idea to recover REEs from phosphogypsum elaborated by **Yamuna Singh** gets significance.

The above efforts for value addition of mining activity have to be complemented by research and innovation in mitigating environmental degradation due to mining. **Barapatre et al.** discussed the utility of microbes for rejuvenating the degraded and contaminated soils in the mining areas. On the similar lines, **Deote et al.** presented a case study of chromium-contaminated soil from Taka area in Central India, wherein the microbes have shown great potential to remediate this soil. The problem of mining and mineral waste is further discussed by **Nikhil Kulkarni**, who emphasized on the sustainable restitution. Another major problem in mining areas is the water quality. **Tumane et al.** discussed the issue of water contamination and toxicity, and proposed several remedial measures. Not only the active mines, but abandoned mines also create several problems. However, such derelict mines can be converted into tourist attraction. This innovative idea is elaborated by **Thakre and Randive** who proposed that the abandoned mines be made tourist attraction for sighting the rare species of bats. The bats have several ecological advantages; therefore, such initiatives can help the conservation of bats as well as bring value addition to mining.

All the authors have taken great efforts in bringing out current state of art in sustainable mining. We have a great pleasure in presenting this book to all the stakeholders.

Kirtikumar Randive
Department of Geology
Rashtrasant Tukadoji Maharaj Nagpur University
Nagpur, India

Shubhangi Pingle
Regional Occupational Health Centre (Southern)
National Institute of Occupational Health
Bengaluru, India

Anupam Agnihotri
Jawaharlal Nehru Aluminium Research
Development and Development Corporation
Ministry of Mines
Government of India
Wadi, Nagpur, India

Acknowledgements

First and foremost, we thank Sherestha Saini, Senior Editor of the Earth and Environmental Science Division of Springer who read and referred our proposal to Alexis Vizcaino. Subsequently, our proposal was received and reviewed by Alexis who was quick to respond, very considerate, patient and ready to help. He helped us throughout the book project, often going beyond his duty as an acquisition editor. We are deeply indebted to him. We are also thankful to our project coordinator Vijaykumar Selvaraj, who kept us motivated and followed-up in such a manner that we never felt exhausted. We are thankful to all reviewers who spared their valuable time for scrutinizing the manuscript and offering comments, criticism and helpful suggestion, which greatly helped for improvement of the manuscript. We are also thankful to all the authors for contributing their valuable research work for this book, and also for their patience and cooperation throughout the book project. We are also thankful to Sanjeevani Jawadand and Sneha Dandekar for the editorial assistance, without their help it would not have been possible to complete this work in time. Finally, we thank Dr. Subhash Chaudhary, Vice Chancellor, RTM Nagpur University, for writing an elaborate forward for this book, and to everyone who is directly or indirectly involved in this work.

Kirtikumar Randive
Shubhangi Pingle
Anupam Agnihotri

Contents

Contributors

Anupam Agnihotri Jawaharlal Nehru Aluminium Research Development and Design Centre (JNARDDC), Nagpur, Maharashtra, India

Gladson Bage Geological Survey of India, Ranchi, India

Sneha Bahadure Jawaharlal Nehru Aluminium Research Development and Design Centre (JNARDDC), Nagpur, Maharashtra, India

Bhagyashree Bangalkar Department of Physics, RTM Nagpur University, Nagpur, India

Anand Barapatre Central Instrumentation Facility, Faculty of Science, Indira Gandhi National Tribal University, Amarkantak, Madhya Pradesh, India

Srinivasa Rao Baswani Regional Petrology Division, Geological Survey of India, Nagpur, India

P. G. Bhukte Jawaharlal Nehru Aluminium Research Development and Design Centre (JNARDDC), Nagpur, (M.S.), India

Nishant Burnase National Institute of Miners' Health, Nagpur, Maharashtra, India

M. J. Chaddha Jawaharlal Nehru Aluminium Research Development and Design Centre (JNARDDC), Nagpur, Maharashtra, India

Mukesh Jitsingh Chaddha Jawaharlal Nehru Aluminium Research Development and Design Centre, Nagpur, India

Sneha Dandekar Metallurgical and Materials Engineering Department, V.N.I.T, Nagpur, India

Reena Das Dr. APJ Abdul Kalam Centre of Excellence in Innovation and Entrepreneurship, Dr. MGR Education and Research Institute, Chennai, India

Basudeb Datta Sierra Mineral Holdings 1 Ltd. (Vimetco), Freetown, Sierra Leone

G. T. Daware Jawaharlal Nehru Aluminium Research Development and Design Centre (JNARDDC), Nagpur, (M.S.), India

Shweta V. Deote Department of Microbiology, Seth Kesarimal Porwal College, Nagpur, Maharashtra, India

Kavita Deshmukh Mayur Industries, Hingna, Nagpur, India

M. L. Dora Regional Petrology Division, Geological Survey of India, Nagpur, India

A. B. Ingle Department of Microbiology, Seth Kesarimal Porwal College, Nagpur, Maharashtra, India

Ruchika Jain National Institute of Miners' Health, Nagpur, India

Shraddha Jaiswal National Institute of Miners' Health, Nagpur, Maharashtra, India

Sanjeevani Jawadand Department of Geology, RTM Nagpur University, Nagpur, India

Aruna Jawade ICMR–Regional Occupational Health Centre (Southern), Bangalore, (KA), India

Amrita Karn Jawaharlal Nehru Aluminium Research Development and Design Centre, Nagpur, India

Manish Kumar Kewat Geological Survey of India, Jabalpur, India

Abhinav Om Kinker Geological Survey of India, Jabalpur, India

Nikhil P. Kulkarni National Institute of Miners' Health, Nagpur, India

Swapnil Magar Department of Microbiology, Seth Kesarimal Porwal College, Nagpur, Maharashtra, India

S. P. Masurkar Jawaharlal Nehru Aluminium Research Development and Design Centre (JNARDDC), Nagpur, (M.S.), India

Rajkumar Meshram Regional Petrology Division, Geological Survey of India, Nagpur, India

Tushar Meshram Regional Petrology Division, Geological Survey of India, Nagpur, India

P. A. Mohamed Najar Jawaharlal Nehru Aluminium Research Development and Design Centre, Nagpur, India

Ashok Nandi Mineral Information & Development Centre (I) Pvt. Ltd., Nagpur, India

Dilip Peshwe Metallurgical and Materials Engineering Department, V.N.I.T, Nagpur, India

Shubhangi Pingle Regional Occupational Health Centre (Southern), NIOH, Bangalore, (KA), India

Suchita Rai Jawaharlal Nehru Aluminium Research Development and Design Centre (JNARDDC), Nagpur, Maharashtra, India

Kirtikumar Randive Department of Geology, Rashtrasant Tukadoji Maharaj Nagpur University, Nagpur, (MH), India

Vishakha Sakhare Jawaharlal Nehru Aluminium Research Development and Design Centre, Nagpur, India

Upendra Singh Jawaharlal Nehru Aluminium Research Development and Design Centre (JNARDDC), Nagpur, (M.S.), India

Yamuna Singh Centre for Earth, Ocean and Atmospheric Sciences, University of Hyderabad, Gachibowli, Hyderabad, India

Hemraj Suryavanshi NHM-II, Geological Survey of India, Nagpur, India

Ajay Kumar Talwar Geological Survey of India, Jabalpur, India

Madhuri Thakare Sant Gadge Maharaj Mahavidyalaya, Hingna, Dist., Nagpur, (MH), India

Rajani Tumane ICMR–National Institute of Occupational Health, Ahmedabad, Gujarat, India

Pooja Zingare Department of Physics, RTMNU, Nagpur, India

A Sustainable Approach to Transforming Mining Waste into Value-Added Products

Sanjeevani Jawadand and Kirtikumar Randive

Abstract Mine waste is usually considered worthless in the subsequent stages of mineral production; but it will become valuable in the future and will be needed for the ever-increasing demand of society. Over several decades, large-scale extraction of minerals to cater ever-increasing demand for the growing global population has generated enormous mining wastes, polluted water bodies and air thereby leaving a deleterious effect on the environment. Each year, several billion tons of solid waste is generated worldwide by mining industries as mine byproducts during all types of mining activities, particularly drilling, blasting, and transportation for the extraction of the desired products. The mining wastes are generated at every step of activity, depending on the type of mining method, type of ore, the geological set-up and processing techniques adopted. It is estimated that about a million ton of ore and waste is generated from the large-scale mines per day, and a couple of thousand tons generates from small-scale mines per day. As per an estimate, mining of 1 ton of coal generates approximately 0.4 tons of waste and tailings. Similarly, the production of 1 ton of copper generates approximately 110 tons of waste and 200 tons removal of overburden. Besides, over 500 million tons of mill tailings is generated per year from various ore concentration processes (e.g., lead-zinc, copper, iron). Therefore, the adoption of optimal waste management strategies is sorely needed for the efficient recycling of large quantities of mining waste products generated each year from hundreds of mining operations across the country (or worldwide). In this study, we present a critical overview of solid wastes in mining and the optimal waste management strategy including legal remedies and economic constraints. A holistic approach to end-to-end mining processes to understand and frame the waste management strategy is needed for reducing the risk and maximizing the resource potential.

Keywords Mining waste · Resource efficiency · Sustainable management

---

S. Jawadand · K. Randive (✉)
Department of Geology, RTM Nagpur University, Nagpur, India
e-mail: randive101@yahoo.co.in

K. Randive et al. (eds.), *Innovations in Sustainable Mining*, Earth and Environmental Sciences Library, https://doi.org/10.1007/978-3-030-73796-2_1

1 Introduction

Mining waste is the result of complex mining processes, which involves steps from exploration to mine closure and the current economic paradigm based on unlimited growth. In each of these steps, a large amount of waste is generated which contain significant levels of toxic substances. So, these mining wastes needs to be tackled effectively to lessen its negative impact on environment. Furthermore, the over-exploitation of mineral resources to meet the burgeoning demands by increasing population result in increased generation of mining waste. The mining waste involves materials that need to be discarded to get access to the mineral resources (such as topsoil, overburden and waste rocks), and materials remaining after selection/processing/treatment (operating residues, as well as tailings); indeed, the desired mineral may be present in the ore in minute amount (less than 1%). It indirectly poses challenge of disposing of such large quantities of waste. Large amounts of overburden and tailings are of growing concern in the mining sector, specifically due to the presence of heavy metals. The storage of the mine wastes is commonly identified as one of the most important causes of environmental impact. The total amount of tailings which would need storage will also surpass the in situ volume of the ore being mined and processed. The global production of solid wastes from primary extraction of minerals and metals is on rise (over 110 billion tons per year) and may range from few times the mass of the valuable element (e.g., iron and aluminium), up to millions of times for scarce elements (e.g., gold) [1].

With the development of mining sector, mining activities have generated more solid waste and caused increasingly serious problems for the environment. An optimal strategy is needed to pursue solid waste management, particularly in the present-day mine waste approach. The mining and processing wastes are considered as a potential resource and have many potential applications because of their low-grade metal content. The waste management system for transforming mining waste into resources creates a complex cluster of inter-related aspects (socio-economic, environmental, and technological) and have many stakeholders including local communities, NGO's, mining and related industries, etc. Inadequate data on mine waste materials impede the identification of most suitable waste management approaches, the optimization of treatment processes, the evaluation of environmental impacts of reuse or landfilling, and the assessment of the release of chemical substances from products or materials. In addition, extracting ores below cut-off grade, recycling mine waste in commercial products and recovering value-based products require advanced processing technologies. Certain mine waste materials contain elements which could be useful to the industry and can be obtained from other sources at a nominal environmental cost. Recycling of the mine wastes extracts valuable resources from the waste stream or transform the entire waste material into a new value-added product [2]. The recycling basically refers to the versatility of waste materials to be processed and transformed into a new product or re-used in almost the same capacity [3].

The diverse forms of mining wastes are of global concern but the present study is limited to solid wastes in mining. Now worldwide, the countries are attempting to adopt the sustainable management strategies for mining waste which further transitions from throwaway society to a zero-waste economy, in which mine waste materials are reused, recycled or reclaimed, and are only disposed of when there is no other alternative. The present study highlights the solid wastes in mining and the holistic approach of end-to-end mining to curb waste and the best use of resources.

2 Mineral Wastes: Nomenclature and Classification

Mineral waste can be defined as a material leftover from exploration, mining and quarrying operation that cannot find a productive use. It is the high-volume material that originates from the excavation and physical and chemical processing of a wide range of metalliferous and non-metalliferous minerals by opencast and deep shaft methods [4]. The mine waste is the geological material below the cut-off grade that is generated during mining operations [5]. Waste rock, tailings, overburden and other solid waste are the largest solid industrial waste produced in the mineral resource exploration process. Much of this waste is used to backfill old pits, create haul roads or bunds, but a lot remains in waste tips or tailings lagoons This increase in mining and mineral waste has serious consequences for humankind, biodiversity and the environment. Different exploitation phases and related solid wastes generated in mining can be classified as follows (Fig. 1).

Waste rock: Waste rock is a bedrock that has been mined and moved out of the pit, but has no commercial metal concentrations [6]. The mine waste rock can further be categorized as clean waste or special waste based on its mineral content and its ability to produce acid [7]. The composition of the waste rock controls the element's behaviour released in the atmosphere.

Tailings: Tailings normally consist of various mixtures of quartz, feldspars, carbonates, oxides, ferromagnesian minerals, and minor amounts of other minerals. Because tailings are essentially finely crushed rocks, their mineralogical composition generally corresponds to that of the parent rock, from which the ore was derived [8]. These are of great concern in the mining sector due to the presence of toxic heavy metals in them. For several mining activities, the storage of tailings is commonly identified as one of the most significant sources of environmental concern. Most probably, the quantity of tailings exceeds the total in situ volume of the mined and processed ore and causes storage issues [9]. For e.g., if the grade of ore were 1% copper, 99% of the total ore would be deposited as tailings. Iron ores generally have higher grades than sulphide or gold ores, often going over 50% or more. Less tailings are therefore produced in the iron ore mines.

Coal Refuse: During the preparation and cleaning of coal, the raw coal is put through a series of shakers to separate rock material from the coal and the resulting solid

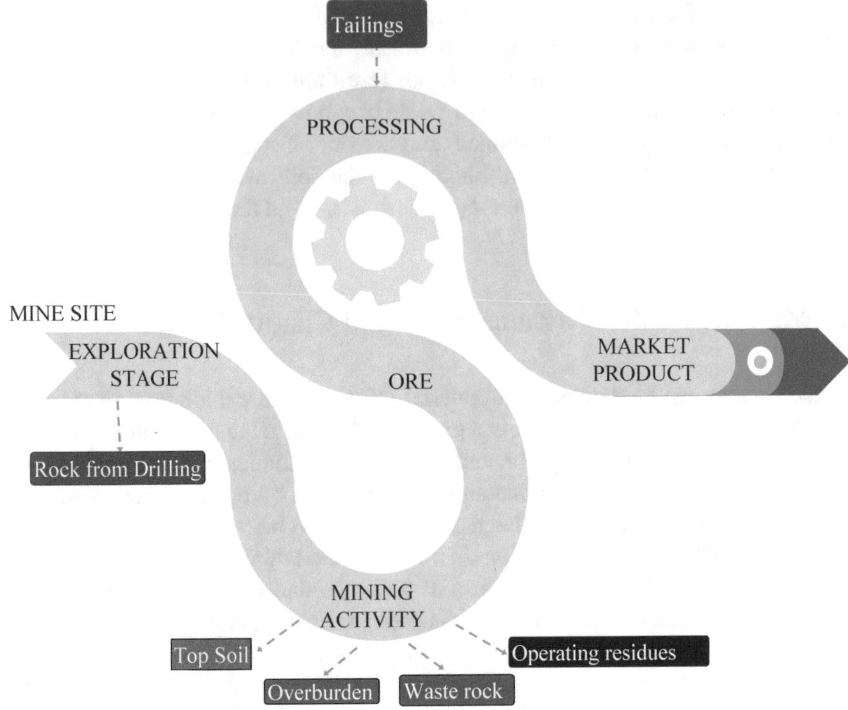

Fig. 1 Solid wastes generated in different stages from exploitation to final product

waste commonly called as coal refuse. Coal refuse contains heavy metals and also a certain amount of sulfur-bearing minerals, especially pyrite and marcasite, that when exposed to water, result in acidic discharge [8]. The processing of 1 ton of hard coal produces 0.4 tons of waste material, much of which is deposited on the ground (about 1300 hectares) and below ground (about 1500 hectares) as waste piles [10, 11]. In India, the opencast mines being a major source of coal production, the ratio of waste generated is much higher than the underground operation due to stripping nature of extraction [12]. In China, about 95% of the total coal output comes from underground coal mines that generate 727.5 million tons of coal mining waste (CMW) annually [13].

Mine dust, aerosols, suspended particles (which settles down to form solid waste): Mining operations such as grinding, milling and management of mine tailings result in coarse particles (about 1 μm diameter) by mechanical action while smelting and refining may result in ultrafine particles (about 0.1 μm) and accumulation mode (0.1–1.0 μm) by condensing high-temperature vapors and subsequent diffusion and coagulation [14–18]. The transport efficiency and deposition of dust particles depend on the particle diameter. The mine dust, aerosols and suspended particles affect the ecosystem health and biogeochemical cycles. It covers leaves and therefore inhibits its capacity to photosynthesis and transpiration, leading to the deterioration of biomass. Secondly increased toxicity in the crops in nearby fields through intake via contaminated soils as well as groundwater. Several effects of mine dust and aerosols from mining operations worldwide are well documented [18].

3 Global Scenario of Solid Wastes in Mining

Owing to the large volume of waste material disposed of from the mines, the problems of mining waste and its management are significant worldwide. In general, an open-pit mine has a higher stripping ratio than an underground mine; which means that generation of waste through open-pit mining is higher. Surface mining operations (open-pit, open-cast or open-cut mining) produce high volume of waste, such as open-pit copper, iron, uranium and taconite mines. For example, the production of 1tonne of copper generates 110 tons of waste and 200 tons of overburden [9].

The global mining sector generates millions of tons of overburden, waste rocks and mineral processing wastes. More than 70% of this global mining waste is generated by UK, Germany, Sweden, Poland and Romania [4]. However, most of these materials end-up being disposed of in landfills either due to their low market value or remote locations of most mining activities [2, 3, 19–21]. At present, almost 100 billion tons of solid waste are generated annually by 3500 mine waste facilities operating worldwide [22, 23]. Mineral processing wastes in the United States accounts for about half of all solid waste produced per year. From 2008 to 2019, Brazil have produced 3.6 billion tons of solid mining waste in dump piles [24]. In India, the Sukinda mining area has produced approximately 7.6 metric tons of solid waste in the form of waste rock and overburden [25, 26]. EU-27 (27 European Union) countries produced about 697 million tons of mining and mineral processing waste in 2016 [27]. In 2018, over two-thirds of the total amount of waste have been generated in EU-27 is major mineral waste [28].

4 Disposal Methods for Solid Wastes from Mines

The proper method of mine waste disposal can circumvent its impacts to some extent. There also needs to build operational flexibility within waste disposal so that waste (material below cut-off grade) will be accessible in the future if commodity prices rise. The concerns related to the disposal of solid waste from metal mines are to choose suitable methods for the systematic use of mining waste and for the management of waste rock and tailing contamination. Some of the solid waste disposal methods are as follows [9].

4.1 Pond Storage

The problems associated with the management of tailings becomes critical with its increasing volume. In pond storage method, the exhausted open pit mines are refilled with the tailings. Technological developments make it possible to mine lower grade ores generating higher volumes of waste that require proper handling. Recently, tailings are stored in mined-out open pits with special designs, especially with uranium mining in Saskatchewan, Canada, so that the transport of pollutants is largely regulated [29]. However, this surface storage may render them vulnerable to potential, sometimes unavoidable disruptions and/or dispersion indefinitely [30]. It can also facilitate percolation of toxic elements into the ground and intoxicate the water table.

4.2 Dry Stacking

In this process, the dewatering of tailings is done using vacuums or pressure filters so the tailings can be stacked [31]. It reduces the potential seepage rates and thereby the impact on the environment. Dry tailing can be transmitted to the tailings storage facility (TSF). However, the construction of TSF is often cost-prohibitive.

4.3 Disposal into Underground Workings

In this process, the disposal of the tailings is done in the exhausted underground mines. It is a somewhat more complex operation than disposal into an exhausted open pit. It involves consideration for stabilization of ground and some type of fill for roof support. So, the tailings need to be combined with additives to bind the water forming a self-hardening fill [32].

4.4 Disposal into the Oceans/Submarine Tailings Disposal (Std)

Tailings can be conveyed using a pipeline and then discharged in such a manner that they can ultimately descend into the depths. In 2015, there were sixteen mining operations around the world using STD [33] as an alternative to land-based tailings storage [29]. Practically, it is not a viable method of reducing the amount of waste as it caused poorly defined environmental impacts.

4.5 Co-disposal of Tailings and Waste Rock

Currently, the methods for the co-disposal of tailings and waste rock to create more stable products and eliminate traditional subaerially discharged slurry tailings are being developed in the mining industry [34]. Thus, depending on the strength and rapid stabilization of the waste co-disposal, the risk and implications of static and dynamic loading have been minimized and allow adequate access to the tailings for rehabilitation [35].

5 Optimal Waste Management Strategy

The mining process generates a significant quantity of waste materials that must be operated strategically to balance economic productivity with environmental sustainability. It will require a holistic approach which integrates all the geological, economic, technical, environmental and social factors. Systematic utilization of solid waste from mines would have considerable social and economic benefits to mitigate resource shortages and for environmental sustainability [36]. Zero waste requires transforming infrastructures and policies, but also education, training, and research. Optimal waste management strategy involves specific approaches as depicted in Fig. 2.

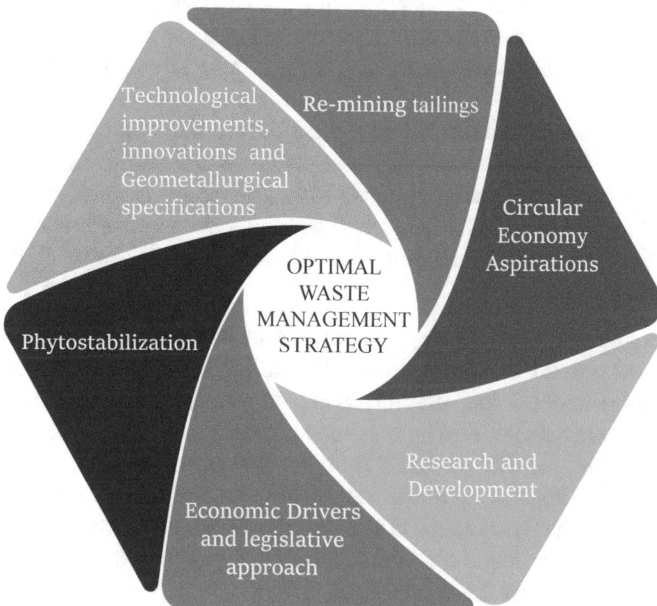

Fig. 2 Waste management strategy

5.1 Technological Innovations

Several technologies are available for management and treatment of mining waste, however, choosing the appropriate one depends on the nature and extent of waste. Characterization and classification play an important role when we need to make a decision on mining waste management. Nevertheless, a regular study in accordance with the application of state-of-the-art technologies is needed to mitigate the severe effects of mining waste and sustain this earth. Some innovative uses of technologies can be explained by the following examples.

The mapping of wastes such as solid waste dumps, tailing ponds, overburden, ore stakes, etc. can help to determine its extent.

The use of hyperspectral imaging in mine waste management and recycling sector provides high performance in terms of material identification. For example, to study the mobility of sediments containing toxic residues [37], the mitigation of toxic metals spread in redox areas [38], the use of biochemical and mineral dissolution processes in sulfurous tailings [39], the dilution of tailings products, the geochemical and mineral elimination of submarine tailings [40], reduction of the discharge of solid waste by recovering sand from ore dressing flow for the construction industry and backfilling approaches [36].

Other remediation systems include the use of heat to volatilize toxic components, and the use of microorganisms to reduce the reactivity and toxicity [41]. There are many instances of successful implementation of microorganisms for the treatment of metal waste in published studies, as well as in ongoing technological applications. For example, Mint Innovation, a start-up in New Zealand, uses micro-organisms to extract metal from waste streams [42]. In this way, we can not only focus on waste treatment but also avoid waste generation altogether by turning potential waste into a resource.

Also, the geochemical characterization establishes the most appropriate form of containment for the post-processing waste materials and improved waste management. Usually some low -grade ore due to lack of advanced technologies considered as a waste and not used in the industry. Technological up-gradation helps effective re-mining the staked low-grade dumps in the mining fields. Some of the success stories are given below.

a. Chromite: Chromite ore having concentrate of >50% Cr_2O_3 with <2.0% SiO_2 which has got a good market potential was earlier considered as a waste. Using advanced technologies, it has become possible to reduce the loss of chromite values in tailings and recover values from previous stacked waste. For example, concentrates with 46–55% Cr_2O_3 can be produced from the low-grade ores with 18–33% Cr_2O_3 by recovering 60–80% of chromite values [43]. In Sukinda, Orissa, the chrome beneficiation plant has been installed to utilize the low-grade chromite ores. Nafziger [44] also describes the major low-grade chromite deposits in the world and gives a review of the recovery methods used.

b. Tungsten: Repositories of historical tungsten mining tailings (e.g., Yxjöberg tungsten mine) are potential resources for valuable metals [45]. Considering the technical viability of reprocessing the tailings, extraction of low-grade tungsten deposits is possible e.g., tungsten from low-grade wolframite deposit.

c. Cryolite: Synthetic cryolite can be recovered as a by-product in phosphate rock processing and alumina manufacture.

d. Nickel: The technology so developed pave the way for the extraction of nickel from lateritic nickel ore and chromite overburden or its utilization in the manufacture of nickel-based chemicals. The secondaries/wastes like catalysts are utilized for the production of high-value metals like nickel, cobalt etc. (e.g., in Institute of Mineral and Materials Technology, Bhubaneswar, National Metallurgical Laboratory NML, Jamshedpur).

e. Iron Ore: Iron-bearing materials are processed for alloy production using different advanced techniques such as smelting reduction, production of advanced materials by adopting chemical/electrochemical/biotechnological routes under different conditions (e.g., in National Institute for Interdisciplinary Science and Technology, NIIST, Thiruvanthapuram).

f. Furthermore, usable minerals could be recycled using advanced mineral processing technologies. The Sivas-Divrigi processing plant in America achieved cobalt recovery ratio of 94.7%, the nickel recovery ratio of 84.6% and copper recovery ratio of 76.8% by recycling of iron tailings using floatation method [36, 46].

5.2 *Phytostabilization*

Phytostabilization is a type of phytoremediation which, by sequestering pollutants in the soil near the roots, uses plants for long-term stabilization and remediation of tailings. Through adsorption or precipitation, plants can immobilize contaminants and provide a zone around the roots wherein the contaminants can accumulate and stabilize. Worldwide, nearly 6.5 million tons of tailings is produced in a single year. So, one of the most common methods of stabilization is by plantation. A phytostabilization approach can be particularly useful in dry environments that are vulnerable to wind and water dispersion. As plantation improves soil moisture content, bulk density, pH and overall soil nutrient content, overburden dumps are typically reclaimed by tree species. The tree species such as *Acacia mangium, Eucalyptus camaldulensis, Dalbergiasissoo, Cassia seamea,* and *Peltophorum pterocarpum* are suitable for bioreclamation of overburden dumps (Fig. 3). By this approach, the contaminants become less bioavailable and the exposure to animals, ecosystems, and human beings is significantly restricted.

Fig. 3 Phytostabilization of mine wastes and ideal plant species

5.3 *Towards the Circular Economy*

As far as the mining sector is concerned, it has been a linear economy model that was in vogue since the 19th century. However, more recently another model called circular economy has been considered seriously by various nations [47]. It is an alternative to a conventional linear economy (make, use and dispose) in which resources are kept for a long-term to extract optimum value from them while in use, and then recover and recycle the products and materials in a synchronised way. Promotion of the 4Rs (Reduce, Reuse, Recycle, Recover) and adoption of the framework of the circular economy (produce-use-return) has created a vibrant new economy in which waste can re-enter the economy in modified form. The introduction of a circular economy model to mining waste presents a huge opportunity to alleviate risk and increase the benefits in terms of value-added products (see Fig. 4). In this transition to sustainability, the material cycle would need to be gradually closed in ways that minimize demand for new minerals and reduce the volume of waste [1]. The espousal of this model for recovering materials as well as energy from waste is not just an environmental obligation but a real economic opportunity. It ensures a resource-efficient pathway to sustainable development. It encourages cradle-to-cradle thinking which emphasizes productive alteration of waste and retrieving it in the system. It supports the concept of upcycling i.e., producing products of greater environmental value than their material inputs over downcycling-based recycling practices which leads to quality loss.

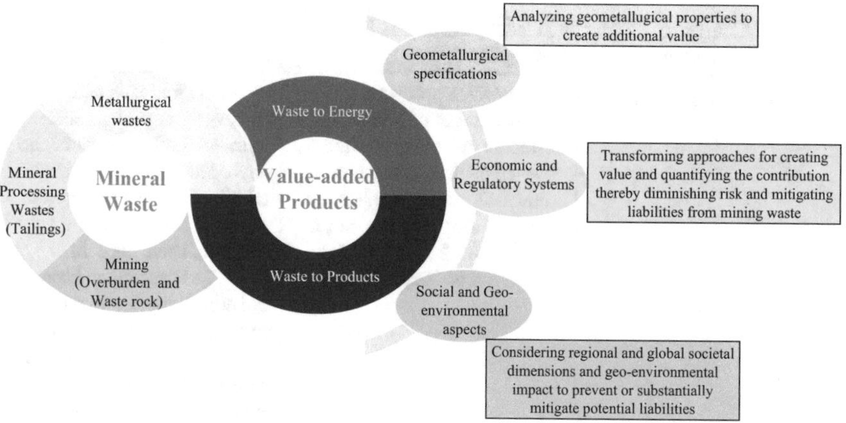

Fig. 4 Towards circular economy—transformation of mineral waste

5.3.1 Some Initiatives and Innovation-Centred Approach

A range of initiatives is taking place throughout the mining industry to improve resource recovery and shift the sector towards a truly circular economy. Several firms are making investments in modern processing technologies to extract and reuse some of the beneficial materials that are often discarded from tons of less usable mining waste, rather than that waste itself [42]. In order to recover as much waste as possible, implementation is done at the mining field and enterprise level itself, in mineral value chain and the system as described by Zhao et al. [48] (Table 1). This circular economy approach in the mining industry in broad scope solves the problems of mineral resource scarcity, growing resource waste and environmental pollution and thus leads to sustainable economic development.

The mining sector has some of the greatest opportunities to reinvent itself in terms of managing waste and converting it into a resource. As the environmental risks associated with extractive waste, both during operation and post-closure are extensive, good planning around these risks requires consideration of climate-related data and information. Through the years, solid waste has evolved in line with technological progress, from multi-centimetre grain size with a still high content of the desired element to micron grain size with very low chemical contents [49]. Research and innovative approach are significant considering the reactivity of specific mining waste. This could be advanced in different ways such as leaching tests, long-term column tests and stabilized tests as being developed in the context of the Landfill Directive. It is advantageous to check the behaviour of metallic molecules (originating from mining waste) in the subjacent geologic layers and within the waste deposit (adsorption and other attenuation processes) and prediction of their fate using tools such as geochemical and solute-transport modelling. Mitsubishi materials with its recycling-oriented business model recycle and recovers metals and also rare metals. JX Nippon Mining & Metals reuses 83% of its total volume of

Table 1 Waste management implementation strategy [48]

Implementation strategy	Description	Example
Individual mining enterprises	Mining waste produced at different stages of processing is treated as raw materials and returned back to initial production stage for other manufacturing processes or as a substitute for raw materials after adequate treatment	Using the ion-exchange process to recycle gallium from the mother liquor of the Bayer alumina factory, the solution was then returned to the main flow, achieving a comprehensive gallium recovery, reducing emissions and enhancing economic performance. Besides, all waste rock can remain in the mine and be used for underground filling using the total tailing cemented fill mining technique, which can effectively release zero discharge
Mining ecological park mode	Cross-linking of different mining and related enterprises to transform waste and energy that cannot be processed further in one mining enterprise into raw material or resource of another enterprise	Ferrous metal mining: Exploration of ore minerals, mineral processing, sintering, steel and iron-making enterprises are cross-linked to other industries and companies to shape an ecological park for ferrous metal mining

waste materials produced in 2015, while its copper recycling system recovers about 26% of its total scrap production [50]. Sumitomo Metal Mining almost doubles the recovery rates of copper scrap in the five years following 2010. Recently, Canada-based Mineworx announced an agreement with Tennessee's Davis Recycling Inc. in April to construct a pilot plant which is a crucial move in demonstrating the effectiveness of the technology; the project will recycle platinum group metals from used catalytic converters [42]. Another firm Comstock Mining, a Nevada-based miner works on improving recovery of mercury from mine tailings and produce over 1,000 tonnes of mercury in the artisanal gold mining sector, owing to the use of mercury in separating gold from non-precious ores [51]. Some firms outside the mining sector (for example, Lafarge, Apple and Tiffany) are focusing on industrial symbiosis and closed-loop production, helping to encourage or drive circularity within it. The L-Max hydrometallurgical method has been developed by the Lepidico company to extract lithium carbonate from lithium mica and phosphate minerals, which is often overlooked by mainstream producers [52]. VTT's collaborative project[1] not only focused on supply chain optimization but the optimal use of residual material from

[1]The project with a collaboration of 19 companies, research organisations, and universities from nine European companies to find ways to improve recovery of a number of waste minerals, such as cobalt, nickel and zinc, and improve Europe's self-sufficiency with regard to metal production.

which metals have already been recovered. The innovation-centred approach gives a better understanding of the barrier and opportunities in the context of mine waste avoidance and resource recovery e.g., MetGrow Calculator.[2]

5.3.2 Enhanced Landfill Mining (ELFM)

Another approach for the productive use of mining waste is Enhanced Landfill Mining (ELFM). Landfill mining is the method by which waste from active or closed landfills is excavated to reduce its environmental effects. It involves extracting the hazardous material from the ground and preparing it for recovery after a predefined time. Waste valorization emphasises the use of any leftover material or by-product to produce other valuable goods and remain as long as possible in production and consumption systems. ELFM includes the valorization of landfill waste, namely waste-to-energy (WtE) and waste-to-material (WtM) in combination with the ecofriendly approach in preventing CO_2 and other pollutants emissions during the valorization processes [53]. It envisages an important major shift in both the waste management vision as well as waste management technology. Its better implementation mostly depends on technological improvements and innovations and surmounting different socioeconomic constraints like social acceptance, protocols, economic uncertainty, etc.

6 Role of Legislations in Management

A set of norms, institutions, and practices are designed to manage waste on a global scale. Rules and guidelines aimed at increasing resource efficiency, reducing waste and encouraging recycling exist around the world. Worldwide, the countries have their mining and environmental legislation which in entire or discretely covers the different concerns of mining activity such as taxes, remunerations, environmental protection, waste management and so on. In most countries, these mining processes and rationales for environment protection are most commonly implemented through a variety of legal tools, such as mining legislation, environmental-related legislation, other legislation and standards, including occupational health and safety. In Canada, laws have been enacted both for the Provinces and at the federal level. In addition to radionuclides that comes under the Atomic Energy Control Bureau, the Canadian Federal Ministry of the Environment has reinforced the regulations on effluents produced by mining activities. Whether in Canada or the United States, the guide values applied at the periphery of the mines are determined according to each type of operation, based on the hydrogeological, physical, chemical and biological properties of the waters receiving the effluents. In Australia, the mining industry is governed by the Australian Mineral Industry Code for Environmental

[2]An online tool that works out the optimal waste recovery process for a mine or region based on characteristic input by the user.

Management, published in 1996. The regulation for large-scale hard rock extraction in Malaysia includes specific criteria for the treatment of tailings. The Mexican Official Standard, approved in 1997, specifies minimum standards for site selection, design, monitoring and maintenance of tailing dams. At the European Union scale, there is currently no clear law on waste from mining activities, nor on the production of raw materials, ore processing or industrial materials. Each of the member States has its mining and environmental legislation. European legislation varies between lateral environmental legislation and legislation in relation to particular industries, products or forms of emissions (Air, water, and waste). In India, the Environment Protection Act of1986 is the central act that pertains to the management of wastes in the country. Moreover, the Management of Hazardous Waste is a complex set of rules which together combine to form the legal regime. Hazardous wastes (HWs) include the wastes generated by various industrial and anthropogenic activities mainly from mining, tailings from pesticide-based agricultural practices, industrial processes of textile, pesticides, tannery, petrochemicals, pharmaceuticals, paints, oil refineries and petroleum processing etc. [54]. With the technological augmentation, Government role in waste management is gradually evolving in response to the changing perceptions in mining operations. Increasingly strict regulations in terms of mining and mineral processing waste, as well as the raised demands for mineral resources, are pushing the mining sector toward higher sustainability to improve cost-effectiveness and meet rising demand. However, the weak institution needs further improvements to constrain mining industries to take strict measures regarding waste management practices. Continuing to tax waste and labor while not taxing resource extraction will impede the transition towards a circular economy.

7 Value Added Products from Mining Wastes

Waste is not the final stage in the life cycle of any object. Landfilling or incinerating puts waste out of sight, but in the end, materials decompose and are transformed into other substances. As a common practice, one specific mine focuses on one specific metal and the rest of the ore is considered waste and disposed of as such. But this toxic waste contains many metals that can be recovered and used sustainably. The mine tailings can be considered for re-mining, instead of being merely discarded. Re-mining tailings extends the operating life of existing mines. The approach of viewing the disused materials from mines (overburden and tailings) as a potential resource rather than waste helps to ascertain strategies for preventing waste or transforming waste into a non-waste resource. Re-visiting mine tailings have more economic and environmental benefits than developing new mines.

An integrated approach to acquire various materials from the same source will make more economic sense [55]. Resources are initially extracted from the environment; but afterwards, with the sustainability approach, the waste itself becomes a resource and it can be re-cycled in the commercial phases. In this perspective, resources need to be managed more efficiently in the entire value chain. Recycling or reuse of mineral waste in commercial products is possible by identifying the potential market and investing in new technologies. The value-added material can also be recovered from solid waste. For example, a green selective enrichment and separation process for the recovery of copper and iron valuable components from Daye copper converter slag [56], utilisation of red mud from Bayer processing of bauxite [57, 58], spent pot lining from aluminium smelting [59], fly ash from power generation, slags from smelting operations and the use of wastes in geopolymer concrete [60].

Considering varying mineral composition in tailings of different mine, following value-added products can be recovered from mine waste:

a. Glass or fertilizer or ceramic products: The use of technogenic raw materials to produce glass, fertilizer or ceramic products provides rational use of mining waste and leads to the introduction of resource-saving technologies. e.g., preparation of glass-ceramics by the iron tailing in Tangshan region or by gold tailings, preparation of soil conditioning agent or the fertilizer by the use of magnetized iron tailings [36, 61–63], use of coal mining waste in ceramic products [64].

b. Construction materials: The use of mining waste materials as concrete aggregates for construction, production of brick and tiles, cement, pozzolana and pigments for paints. Modular bricks can be prepared using iron ore waste with fly ash and cement as aggregates. Chen et al. [65] carried out studies on utilization of hematite tailings in production of non-fired bricks. Niu et al. [66] reveals that the concrete products obtained by mixing the iron ore tailings, cement and fly ash at a ratio 65:25:10. Dean et al. [67] used the gold mill tailings in addition to fly ash, Portland cement and water to manufacture concrete blocks of size 10.16×20.32 cm. Roy et al. [68] carried out an experimental study on gold mill tailings of Kolar Gold Fields in the making of bricks. Anglo American used coal mining waste in Mpumalanga province, South Africa to turn mining waste (gypsum) into bricks [69].

c. Technical-artistic material: Development of innovative polymer-based composite materials, extracted from non-contaminated waste-rock tailings which then used for creative value-added applications including renovation, reconstruction of historic monuments, artwork, decorative and architectural interventions, or simply as coating materials. e.g., reuse of rock-waste tailings, a by-product from Panasqueira mining operations in Central Portugal, one of the largest tungsten mines in the world [70].

d. Compact composites: Other added-value applications for reusing mining and quarrying wastes are the production of compact composites, particularly, from marble or quartz wastes [70]. These mining by-products could also be used for 3-D printing. For e.g., The Additive Manufacturing arm of Sandvik uses diamond composites in 3-D printing, and the 3-D printing pump components of Markforged Industries work on a range of items that use different metals [52]. The economic value of such composites depends primarily on its aesthetic appeal, which is imparted primarily by the unique textures and colour scales of the wastes.

8 Merits of Mining Waste Transformation

Mining waste transformation ensures the sustainable management of mining waste. It can be substantial in sustaining industrial ecology. It is a broader concept that adds additional economic and social sustainability dimensions to that of environmental sustainability. In this consumption-driven society, where the large volume of solid wastes generated during mining and mineral processing cause irresponsible continuous environmental impacts and affect the global ecosystems. Finding new markets and applications in other sectors of the economy, recycling and re-use of the different types of wastes is a potential panacea to the environmental and health challenges posed by the enormous mine waste. This can minimize mineral waste with careful consideration and optimization of the processing plants. Besides, it minimizes the necessity of extracting the initial resources and provides additional benefits such as energy savings, and pollution reduction. An approach to mining waste transformation integrates actions targeted at transforming waste into a non-waste resource and reducing negative environmental impacts, taking into account economic efficiency and productivity [71]. Efficient use of waste as a resource proves to be an efficient tool for minimizing environmental burden. Strategies and action-oriented programs may help in the "transition" from waste to resource. It helps in producing commercial assets, increasing resource efficiency, reducing waste production and accumulation, generating employment and shared responsibility for the environment. It also lessens the amount of land usage and reduces the impacts on eco-environment.

9 Summary

Mining waste is hazardous to human beings and for the environment too, so it is necessary to choose proper methods for the disposal and management of mining waste. With the general perception, that mine waste management is a burden; the present paper has elaborated the approach that could maximize resource utilization and value generation from waste. The sustainable management strategies for mining waste adopts transition from throwaway culture to a circular economy. To better implementation, the comprehensive framework and optimal waste management strategy play an integral part that comprises all sustainability dimensions. Ultimately, national and local laws dictate which waste disposal option is worthy of consideration. Innovations, along with existing technologies, include environmentally and economically viable solutions to put the mining industry ever closer to its zero-waste objective. Reprocessing waste material is a new commercial opportunity for mining firms to expand their production system and explore potential markets. It will not only help mining firms to regulate their resources and investments in the long term, but it will also secure a new stream of revenue for the miner and support industrial activities.

References

1. Rankin J (2015) Towards zero waste. Aus IMM Bull 32–37
2. Lottermoser BG (2011) Recycling, reuse and rehabilitation of mine wastes. Elements 7:405–410
3. Ndlovu S, Simate GS, Matinde E (2017) Waste production and utilization in the metal extraction industries. Taylor & Francis/CRC Press, Boca Raton, FL. https://doi.org/10.1201/9781315153896
4. Szczepańska J, Twardowska I (2004) III.6−mining waste. In: solid waste: assessment, monitoring and remediation. Waste Manag Ser 4:319–385
5. Scoble M, Klein B, Dunbar WS (2003) Mining waste: transforming mining systems for waste management. Int J Surf Mining Reclam Environ 17(2):123–135
6. Backforty mine (2018) What is the difference between waste rock and tailings?. http://backfortymine.com/2018/08/30/the-difference-between-waste-rock-and-tailings/
7. Impact Assessment agency of Canada (2016) Panel report. 6.0 mine waste rock management. https://www.ceaa.gc.ca/default.asp?lang=En&n=29CBBFF81&offset=7&toc=show
8. Collins RJ, Miller RH (1979) Utilization of Mining and Mineral Processing Wastes in the United States. Miner Environ 1:8–19
9. Das R, Choudhury I (2013) Waste management in mining industry. Indian J Sci Res 4(2):139–142
10. Fecko P, Tora B, Tod M (2013) Coal waste: handling, pollution impacts and utilization. In: The coal handbook: towards cleaner production coal utilisation, vol. 2. Woodhead Publishing Series in Energy, pp 63–84. https://doi.org/10.1016/B978-1-78242-116-0.50020-7
11. Pietrzykowski M, Krzaklewski W (2018) Reclamation of mine lands in poland: biogeotechnologies for mine site rehabilitation
12. Bishwal RM, Sen P, Jawed M (2019) Future challenges of overburden waste management in indian coal mines. In: Ghosh SK (ed) Waste Management and Resource Efficiency. In: Proceedings of 6th IconSWM 2016. Springer Nature Singapore Pt Ltd., pp 1003–1011. https://doi.org/10.1007/978-981-10-7290-1
13. Cui L, Cheng F, Zhou J (2015) Behaviors and mechanism of iron extraction from chloride solutions using undiluted Cyphos IL 101. Ind Eng Chem Res 54:7534–7542
14. Jacob DJ (1999) Introduction to atmospheric chemistry. Princeton University Press, Princeton, NJ
15. Banic C, Leaitch WR, Strawbridge K et al (2006) The physical and chemical evolution of aerosols in smelter and power plant plumes: an airborne study. Geochem: Explor Environ, Anal 6:111–120
16. Wong HKT, Banic CM, Robert S et al (2006) In-stack and in-plume characterization of particulate metals emitted from a copper smelter. Geochem: Explor Environ, Anal 6:131–137
17. Zdanowicz CM, Banic CM, Paktunc DA (2006) Metal emissions from a Cu smelter, Rouyn-Noranda, Québec: characterization of particles sampled in air and snow. Geochem: Explor Environ, Anal 6:147–162
18. Csavina J, Landazuri A, Wonaschutz A et al (2011) Metal and metalloid contaminants in atmospheric aerosols from mining operations. Water Air Soil Pollut 221(1–4):145–157. https://doi.org/10.1007/s11270-011-0777-x
19. Bian Z, Miao X, Lei S et al (2012) The challenges of recycling mining and mineral processing wastes. Science 337(6095):702–703
20. Flanagán C, Grail BM, Johnson DB (2017) New approaches for extracting and recovering metals from mine tailings. Miner Eng 106:71–78
21. Matinde E, Simate GS, Ndlovu S (2018) Mining and metallurgical wastes: a review of recycling and re-use practices. J Southern African Inst Mining Metall 118(8): 825–844. https://dx.doi.org/10.17159/2411-9717/2018/v118n8a5
22. Tayebi-Khorami M, Edraki M, Corder G, Golev A (2019) Re-thinking mining waste through an integrative approach led by circular economy aspirations. Minerals 9(5):286. https://doi.org/10.3390/min9050286

23. Nishant Singh, Pankaj Kumar, Nitin Singh (2020) Geotechnical evaluation of mining waste by utilising. Int Res J Eng Technol (IRJET) 7(5):7757–7760
24. Carmo FF, Lanchotti AO, Kamino LHY (2020) Mining waste challenges: environmental risks of gigatons of mud, dust and sediment in megadiverse regions in Brazil. Sustainability 12:8466. https://doi.org/10.3390/su12208466
25. Mohanty M, Patra HK (2011) Attenuation of chromium toxicity in mine waste water hyacinth. J Stress Physiol Biochem 7(4):336–346
26. Tiwary RK, Dhakate R, Rao VA, Singh VS (2005) Assessment and prediction of contaminant migration in ground water from chromite waste dump. Environ Geol 48:420–429
27. Kulczycka J, Dziobek E, Szmiłyk A (2020) Challenges in the management of data on extractive waste—the Polish case. Miner Econ 33:341–347. https://doi.org/10.1007/s13563-019-00203-5
28. Eurostat (2020) Waste statistics. https://ec.europa.eu/eurostat/statistics-explained/index.php?title=Waste_statistics&oldid=504290#Total_waste_generation
29. Kwong YTJ, Apte SC, Asmund G et al (2019) Comparison of environmental impacts of deep-sea tailings placement versus on-land disposal. Water Air Soil Pollut 230:287. https://doi.org/10.1007/s11270-019-4336-1
30. Kilborn Ltd and Beak Consultants Ltd (1979) An assessment of the long-term suitability of present and proposed methods for the management of uranium tailings. Atomic Energy Control Board, INFO-0024, Ottawa, Ontario
31. Davies MP, Rice S (2001) An alternative to conventional tailing management-"dry stack" filtered tailings. In: Proceeding of the eighth international conference on tailings and mine waste. Fort Collins, Colarado, US: Balkema, 411–422
32. AECB (1995) Report: assessment of the Underground disposal of tailings. Atomic Energy Control Board Canada. AECB Project no. 5.147.1
33. GESAMP (Joint Group of Experts on the Scientific Aspects of Marine Environment Protection) (2016) GESAMP reports & studies No 94—Proceedings of the GESAMP international workshop on the impacts of mine tailings in the marine environment, June 10–12, 2015, Lima, Peru, p 83
34. Mining Magazine (2017) Goldcorp trials new EcoTails technology, p 30. https://www.miningmagazine.com/future-of-mining/future-of-mining-sustainability/goldcorp-trials-new-ecotails-technology/
35. DPI (2003) Management of tailings storage facilities-environmental guidelines. Department of Primary Industries, Victoria
36. Lu Z, Cai M (2012) Disposal methods on solid wastes from mines in the transition from open-pit to underground mining. Proced Environ Sci 16:715–721
37. Pattelli G, Rimondi V, Benvenuti M et al (2014) Effects of the november 2012 flood event on the mobilization of hg from the mount amiata mining district to the sediments of the paglia river basin. Minerals 4:241–256
38. Lynch SFL, Batty LC, Byrne P (2014) Environmental risk of metal mining contaminated riverbank sediment at redox-transitional zones. Minerals 4:52–73
39. Nordstrom DK (2011) Mine waters: acidic to circumneutral. Elements 7:393–398
40. Dold B (2014) Submarine tailings disposal (STD)—A review. Minerals 4:642–666
41. Johnson DB (2014) Recent developments in microbiological approaches for securing mine wastes and for recovering metals from mine waters. Minerals 4:279–292
42. Casey JP (2020) Circular economy: the projects leading the way in mining waste recovery. Mining Technol. https://www.mining-technology.com/features/circular-economy-the-projects-leading-the-way-in-mining-waste-recovery/
43. Reddy PSR, Das B, Rao RB, Misra VN (2004) Utilization of Low-grade chromite ores of Orissa. In: Rao GV, Misra VN (eds) Mineral Processing Technology. Allied Publishers, pp 497–507
44. Nafziger RP (1982) A review of the deposits and beneficiation of lower-grade chromite. J South African Inst Mining Metall 205–226
45. Munlenshi J, Khavari P, Chelgani SC, Rosenkranz J (2019) Characterization and beneficiation options for tungsten recovery from yxsjoberg historical ore tailings. Processes 7(895):19p. https://doi.org/10.3390/pr7120895

46. Sirkeci AA, Gul A, Bulut G (2006) Recovery of Co, Ni, and Cu from the tailings of divrigi iron ore concentrator. Miner Process Extra Metall Rev 27(2):131–141
47. Dandekar SN, Jawadand SA, Dora ML et al (2017) Challenges and prospects in implementing zero waste mining in contemporary scenario. In: Conference International Conference and Expo on Mining Industry Vision 2030 & Beyond, pp 134–141
48. Zhao Y, Zang L, Li Z, Qin J (2012) Discussion on the model of mining circular economy. Int Confer Future Energy Environ Mater Energy Proced 16:438–443
49. BRGM (2001) Management of mining, quarrying and ore-processing waste in the European Union, 79: p 7 Figs., 17 Tables, 7 annexes, 1 CD-ROM (Collected data)
50. ICMM (2020) The 'circular economy' in mining and metals. https://miningwithprinciples.com/the-circular-economy-in-mining-and-metals/
51. Esdaile LJ, Chalker JM (2018) The mercury problem in artisanal and small-scale gold mining. Chem Eur J 24:6905–6916. https://doi.org/10.1002/chem.201704840
52. Leonida C (2020) Turning mine waste from risk into opportunity. The Intell Miner. https://theintelligentminer.com/2020/06/10/turning-mine-waste-from-risk-into-opportunity/
53. Abdel-Shafy HI, Mansour MSM (2018) Solid waste issue: sources, composition, disposal, recycling, and valorization. Egyp J Petrol 27(4):1275–1290. https://doi.org/10.1016/j.ejpe.2018.07.003
54. Mane AV (2014) A critical overview of legal profile on solid waste management in India. Int J Res Chem Environ 5(1):1–16
55. Commission Planning (2007) Report of working group on minerals exploration and development (other than coal and lignite) for the eleventh five year plan. Govt. of India, New Delhi
56. Cao H, Wanga J, Zhang L, Sui Z (2012) Study on green enrichment and separation of copper and iron components from copper converter slag. Proced Env Sci 16:740–748
57. Jahanshahi S, Bruckard WJ, Somerville MA (2007) Towards zero waste and sustainable resource processing. In: International conference on processing and disposal of mineral and industry Waste (PDMIW'07), p 1, Falmouth, UK, 14–15
58. Alam MK, Zanganeh J, Moghtaderi B (2018) The composition, recycling and utilisation of Bayer red mud. Res Conserv Recycl 141:483–498. https://doi.org/10.1016/j.resconrec.2018.11.006
59. Mansfield K, Swayn G, Harpley J (2002) The spent pot lining treatment and fluoride recycling project. In: Green processing 2002: 307, australian institute of mining and metallurgy. Melbourne
60. Davidovits J (2008) Geopolymer chemistry and applications. Institut Géopolymère, Geopolymer Institute, Saint-Quentin, France. https://www.researchgate.net/publication/265076752_Geopolymer_Chemistry_and_Applications
61. Zhang J, Wen Ni, Wang Y (2005) Research on producing glass-ceramics by iron tailings. Metal Mine Chinese 353(11):72–74
62. Xing J, Song S, Xu X (2001) Preparation of gold tailings glass-ceramics. Chinese J Nonferrous Metals 11(2):319–322
63. Wang Y, Chang Q (1999) The present conditions of China's iron tailings resource and the art-of-state and effort direction of its utilization. Metal Mine Chinese 271(1):1–6
64. Lemeshev VG, Gubin IK, Savel'ev YA et al (2004) Utilization of coal- mining waste in the production of building ceramic materials. Glass Ceram 61:308–311
65. Chen Y, Zang Y, Chen T, Liu T (2011) Utilisation of hematite tailings in non-fired bricks production. In: Fifth International Conference on Informatics and Biomedical Engineering (ICBE), pp 1–4
66. Niu X, Chen J (2011) Effect of additives on the properties of concrete products made from iron ore tailings. In: ISWREP 2011−proceedings of 2011 international symposium on water resource and environmental protection, vol 3. https://doi.org/10.1109/iswrep.2011.5893361
67. Dean JD, Stephens JE, Bancroft J (1996) Use of mine tailings in construction materials. Tailings and mine waste. Balkema, Rotterdam, pp 567–575

68. Roy S, Adhikari GR, Gupta RN (2007) Use of gold mill tailings in making bricks: a feasibility study. Waste Manag Res 25(5):475–482
69. Cheeseman GM (2011) Coal mining waste used to build houses in South Africa. Triple Pundit. https://www.triplepundit.com/2011/06/coal-mining-waste-make-d
70. Castro-Gomes JP, Silvab AP, Canoc RP et al (2012) Potential for reuse of tungsten mining waste-rock in technical-artistic value-added products. J Cleaner Prod 25:34–41
71. Randive KR, Jawadand SA (2020) Mineral economics: an indian perspective. Nova Science Publishers, New York, pp 213–246

Utilization of Aluminium Industry Solid Waste (Red Mud/Bauxite Residue) in Pollution Control

Suchita Rai, Sneha Bahadure, M. J. Chaddha, and Anupam Agnihotri

Abstract In developing countries like India, constantly increasing industrialization leads to economic growth but also adds to solid and liquid waste generation. In many developed and developing countries waste disposal is becoming an ever-increasing problem. Improper management of the pollutants in the atmosphere cause numerous environmental problems. Red mud/bauxite residue is one of the industrial solid wastes generated during the extraction of alumina from bauxite in an alumina refinery. It contains undigested alumina, oxides of iron, titanium and sodium alumino silicates along with caustic soda, the latter in the adherent liquid phase. Due to the relevant adsorption properties, amended/neutralized red mud has attracted researchers to work upon the utilization of red mud in adsorption of toxic elements and compounds such as oxides of nitrogen and sulphur from flue gases; heavy metal ions, inorganic anions (fluoride, nitrate and phosphate), dyes, phenols and its derivatives from industrial effluents. This paper is an effort to compile the data on the work carried out nationally and internationally from 1964 to 2019 in pollution control using red mud. Chemical composition of red muds generated in different Indian alumina refineries has been given in the paper. Possible applications of red mud in distinct areas of pollution control have been discussed thoroughly and have been tabulated elaborately. Issues which are restricting commercialization of this application of red mud have also been focused upon.

Keywords Red mud/bauxite residue · Pollution control · Adsorption · Industrial effluents

S. Rai (✉) · S. Bahadure · M. J. Chaddha · A. Agnihotri
Jawaharlal Nehru Aluminium Research Development and Design Centre (JNARDDC), Wadi, Amravati Road, Nagpur 440 023, Maharashtra, India
e-mail: suchitarai@jnarddc.gov.in

M. J. Chaddha
e-mail: mjchaddha@jnarddc.gov.in

A. Agnihotri
e-mail: director@jnarddc.gov.in

K. Randive et al. (eds.), *Innovations in Sustainable Mining*, Earth and Environmental Sciences Library, https://doi.org/10.1007/978-3-030-73796-2_2

1 Introduction

Industries are proved to be the prominent factor responsible for the GDP (Gross Domestic Product) of any country in terms of income, employment, productivity, economical and agricultural growth. But industrialization is in turn directly proportional to the waste generation. A few of them may be in the form of solids (fly ash, goethite sludge, red mud), liquids (industrial effluents) or gases (flue gases) which impart adverse effects on the environment and mankind. Industrial waste generation in the production of goods is inevitable. Merely disposing the waste in the dumping sites increases the severity of the problem which may become unmanageable. This leads to severe consequences such as environmental hazards, health hazards, unpleasant surroundings and land shortage. Hence a need arises for sustainable management of these generated wastes. Proper segregation and scientific recycling/utilization of each component present in the waste may be the solution for this problem which can be seen from Fig. 1 showing the waste management hierarchy.

In the Bayer process, alumina is extracted from the bauxite ore generating a waste product called as Red mud/bauxite residue. Yearly, global generation of bauxite residue is about 150 million tonnes [1]. India contributes about 6% to the total generation of red mud [2]. During the alumina production, bauxite is reacted with caustic soda at elevated temperature (106–240 °C) depending upon the bauxite minerology being processed. Hence it is alkaline in nature with pH ranging from 12–13.5. Typical chemical composition of Indian red mud is given in Table 1 while the typical mineralogical phases present are given in Table 2. Mineralogy of the red mud primarily depends upon the mineral phases of the bauxite and the technology being used. Red mud is generally disposed of in landfills which acquire a large surface of land. Groundwater contamination is one of the environmental problems due to alkali

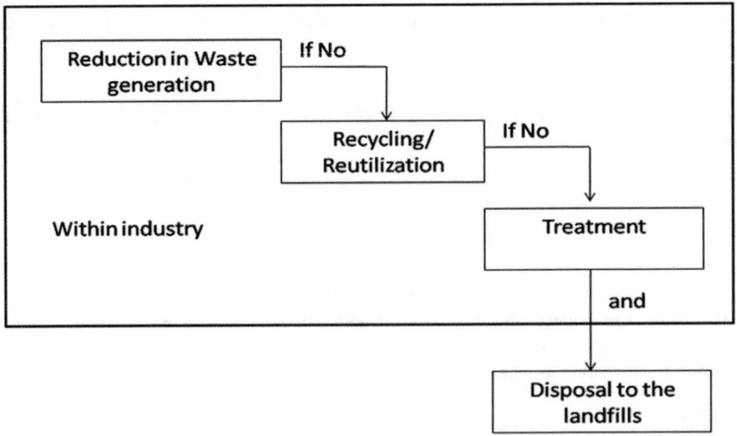

Fig. 1 Waste management hierarchy

Table 1 Typical composition of Indian red mud

Composition, %	Weight %
Al_2O_3	16–23
Fe_2O_3	40–62
SiO_2	4.5–9.5
TiO_2	4–16
CaO	0.5–3
Na_2O	3–6.5
LOI	8.5–13

Source Data collected from various alumina refineries in India

Table 2 Typical mineralogical phases in Indian red mud

Alumina as	Gibbsite
	Boehmite
	Alumo-goethite
	Bsodalite
	Kaolinite
	Diaspore
Silica as	Kaolinite
	Bsodalite
	Quartz
Titania as	Rutile
	Anatase
	Ilmenite
Iron as	Hematite
	Alumo-goethite
	Ilmenite
Na_2O as	Bsodalite
CaO as	Calcite

seepage from red mud. Red mud also impacts on plant life due to alkali dust deposition on them. Hence it has become a universal problem for the alumina producers. Therefore, investigations are being carried out by the researchers in utilization of red mud in applications such as building and construction, metal recovery, catalyst and catalyst support, vegetation, soil stabilization, paints and pigments and pollution control.

As seen from Table 1, red mud/bauxite residue consists of oxides of aluminium, iron, titanium, calcium, and sodium aluminosilicates. Metal oxides are favourable for the fluoride remediation [3]. Also, iron oxide helps in adsorption of heavy metals, anions and hazardous elements in wastewater. Titanium dioxide is found to be potential adsorbent for halogens, fluoride ions and arsenic compounds [3]. Red mud has a very fine size ($d_{50} < 10\ \mu$) and BET surface area 14–25 m^2/gm. Red mud has these adsorption characteristics and can be utilized for the adsorption of certain elements from waste effluents and flue gases. This paper is an overview on the work carried out on global scale including India on red mud utilization in pollution control. Identified issues and research gaps related to its commercialization in this application area has been explained briefly.

2 Utilization of Red Mud in Pollution Control

Pollution control through red mud is the major area of interest for the researchers all over the world. Oxides of sulphur and nitrogen are present in flue gases. Inorganic anions, metal ions, phenols and its derivatives, dyes are present in the industrial effluents as well as wastewater. The sulphur and nitrogen oxides can be oxidised in the atmosphere to form sulphuric and nitric acids resulting in the production of acid rain [4]. Heavy metals present in the industrial wastewater can cause serious health hazards such as reduced growth and development, malfunctioning of the nervous system, organ damage and death in the extreme cases [5]. Other pollutants such as dyes present in the wastewater of the textile industry negatively impact the aquatic life. It reduces the light penetration for photosynthesis in bacteria and plants. They also result in the addition of aromatics, chloride, metals and color increasing toxicity levels in the aquatic ecosystem [6]. These toxic pollutants can be adsorbed on red mud as it possesses the requisite properties or can be modified by treatment for adsorption. Figure 2 shows applications of red mud as adsorbents in pollution control. Table 3 shows some of the industrial effluents and their sources.

2.1 Oxides of Sulphur, Nitrogen Removal

SO_2 is soluble in water and found to be in flue gases of sulphuric acid plant, pulp and paper industry, production of elemental sulphur, and fossil fuel combustion. Nitrogen is present in industrial flue gases in the form of nitric oxide (NO) and nitrogen dioxide (NO_2) [7]. Flue gases from thermal power plants, nitric acid manufacture, explosives industry and welding processes are the major source of oxides of nitrogen. Several studies have been conducted for the adsorption of SO_x and NO_x on red mud.

A process was patented for the first time in France and Japan to absorb sulphur dioxide (SO_2) in industrial effluent gases [8]. A study was carried out in which along with limestone, dolomite and chalk, activated red mud was also used to remove SO_2 from pilot moving grate furnace stack. A laboratory scale bubbling reactor with a continuous feed of both gas and liquid phases was used to study the ability of red mud for adsorption of SO_2. A pilot plant study was carried out for sulphur removal in electric power plant gas using red mud. Similarly use of red mud has been investigated for removal of oxides of nitrogen from flue gases. A solid catalyst was developed

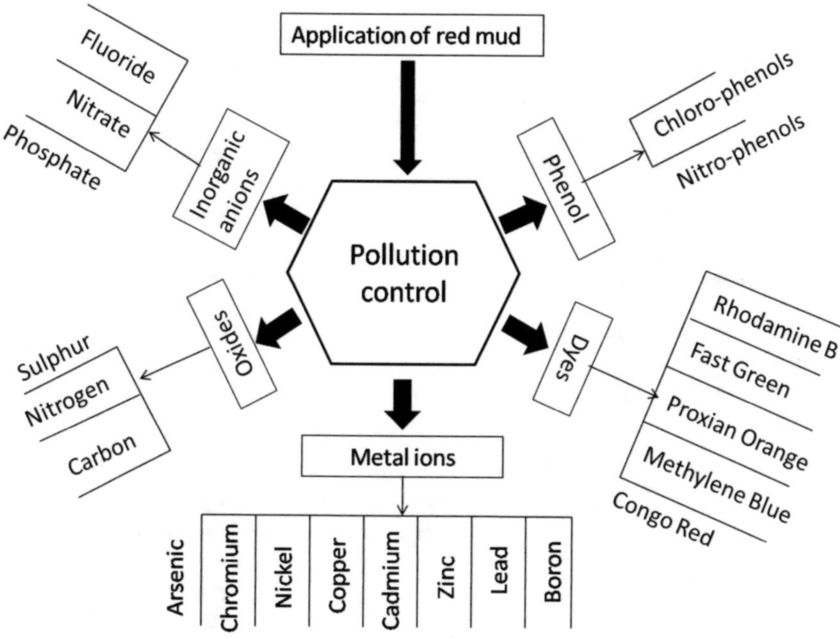

Fig. 2 Applications of red mud as an adsorbent in pollution control

Table 3 Industrial effluents and their sources

Industrial effluent	Source	Refs.
Inorganic anions	Superphosphate fertilizer industry, oil refineries, explosive industry	[3]
Oxides of sulphur and nitrogen	Manufacturing of sulphuric acid, Burning of fossil fuels	[7]
Toxic heavy metal ions	Electroplating and metal surface treatment processes	[8]
Dyes	Textile industry	[4]
Phenols and its derivatives	Oil refining, petrochemicals, pharmaceuticals	[9]

from red mud for treating oxides of sulphur and nitrogen in waste gases. A process was patented by Sumitomo in which red mud was used for the adsorption of SO_2 from exhaust gases. The studies carried out for adsorption of oxides of sulphur and nitrogen on red mud has been tabulated in Table 4.

Table 4 Adsorption of oxides of sulphur and nitrogen on red mud

Sr. no.	Experimental conditions				Country	Remarks	Refs.
	Adsorbent	Waste effluent	Adsorbate	Operating conditions			
1	40% aqueous suspension of red mud	Industrial effluent gases	SO_2	pH 4.3	France, Japan	$NaHSO_3$ solution was filtered off and recovered	[8]
2	Limestone, dolomite, chalk, and red mud	Grate furnace stack gas	SO_2	–	Columbia	10–65% SO_2 removal was observed	[11]
3	Activated red mud	Stack gases	SO_2	500 °C	Japan	Complete removal i.e. 1.8% present in stack gases at 500 °C	[12]
4	Treated red mud pellets	Waste gases	SO_2, NO_x	Catalytic adsorption	Japan	Red mud acted as an adsorbent for SO_2 and reductant for NO_x	[13]
5	Red mud	Exhaust gases	SO_2	Cooling from 100 °C to room temp	Germany	About 97% SO_2 absorption efficiency was observed	[14]
6	Red mud suspension	Synthesized gas	SO_2	Laboratory scale bubbling reactor	Italy	High removal efficiencies can be achieved in the bubbling reactor	[15]

2.2 *Toxic Heavy Metal Ions Removal*

Heavy metals such as Cd, Cr, Cu, As, Ni, Zn, and Pb are the hazardous metals present in the industrial effluents such as in electroplating and metal surface treatment process plants [4]. Acid treated, modified, neutralized, thermally treated, activated red mud has been studied for the adsorption of arsenic from waste effluent. Ferrous based red mud sludge (FRS) was used for arsenic removal from aqueous solution. Arsenic removal was also studied by using polymeric/inorganic sorbents synthesized from red mud and nano-size magnetite. Activated red mud was used to adsorb phosphate and heavy metal ions present in waste water. Also use of de-alkaline red mud has been investigated for removing heavy metals from crude oil and their derivatives like asphalt, tar, sand oils or shade oils. Granular, bacteria modified, and iron oxide activated red mud were examined as low-cost adsorbent for its ability to adsorb cadmium from aqueous solution. Nickel (Ni^{2+}) ions from waste water could be removed by using washed Bosnian red mud as an adsorbent and was found economical. Similar studies were carried out with acid treated red mud. Batch adsorption study was carried out for the adsorption of zinc ions (Zn (II)) from the aqueous solutions with red mud which was activated and neutralized using CO_2. The investigations carried out for adsorption of heavy metals using red mud is shown in Table 5.

2.3 *Inorganic Anions Removal*

Inorganic anions such as fluorides, phosphates and nitrates are found in industrial waste water streams. Fluorides are mainly occurred in waste effluents of steel industry, thermal power plants, aluminium and zinc smelters, superphosphate fertilizer industry, glass and ceramic manufacturing processes, oil refineries, beryllium extraction plants, solar cell industry, silicon based high-tech semiconductors and in municipal waste incineration plants [3]. Explosive industry wastewater is the source of nitrate and organic nitro compounds.

Acid treated/neutralized red mud (20% HCl) was used for the removal of phosphate at different heating and drying conditions. Similar study was carried out for the removal of entrained solids and dissolved phosphates with the aid of treated red mud (1:1 of red mud and 10% sulphuric acid). Raw, activated and granular red mud have been used for removal of fluoride and nitrate from aqueous solution. Investigations conducted for the removal of inorganic anions have been tabulated in Table 6.

2.4 *Dyes Removal*

Textile industry is the major source of wastewater containing dyes such as rhodamine B, fast green, methylene blue, congo red, and procion orange [9]. Acid treated red mud was used as a sorbent for the removal of cesium-137 and strontium-90 from water. Red mud was used for removal of congo red, acid violet, and procion orange from wastewater. Similarly, researchers have worked upon using raw and activated red mud for removal of dyes such as rhodamine B, fast green, methylene blue and remazol brilliant blue dye, malachite green (MG) and crystal violet (CV) from wastewater. Microwave activation method was employed to activate red mud which was ultimately used for the removal scatter blue E-4R from dye wastewater. Along with red mud other adsorbents such as Fe- impregnated clay, Portuguese clay were used for the removal of Acid-orange 7. A detailed overview and experimental conditions for the adsorption of dyes on red mud has been given in Table 7.

Table 5 Adsorption of toxic metal ions on red mud

Sr. no.	Experimental conditions				Country	Remarks	Refs.
	Adsorbent	Waste effluent	Adsorbate	Operating conditions			
1	Activated red mud	Coal	Phosphorous and other heavy metals	400–800 °C, 30 min	China	Removed phosphorous from waste water effectively at pH 5.7. Zn, Cu, Cd were also removed	[16]
2	De-alkaline red mud	Crude oil and its distillates	Heavy metals	350–500 °C	Japan	De-alkalized/neutralised red mud was found more effective than alkaline red mud	[17]
3	Activated red mud	Aqueous synthetic solution and industrial effluents	Chromium	Gentle stirring	India	Significant adsorption was observed at pH 5.2 and a temperature of 30 °C	[18]
4	Neutralized red mud	Waste effluents	Arsenic	80050 cycle/min	Turkey	Activated red mud can be used economically for arsenic adsorption	[19]
5	Neutralized red mud	Wastewater	Arsenate	190 strokes/min	Denmark	Efficient low cost adsorbent for pre-treating arsenate contaminated water	[20]
6	Activated seawater neutralized red mud	Wastewater	Inorganic arsenic	23 °C, 48 h	Australia	As(III) needs to be oxidized to As(V) for favourable removal using red mud	[21]

(continued)

Table 5 (continued)

Sr. no.	Experimental conditions				Country	Remarks	Refs.
	Adsorbent	Waste effluent	Adsorbate	Operating conditions			
7	Neutralized red mud	Aqueous solution	Boron	500 rpm in room temp	Turkey	Langmuir and Freundlich isotherm models were studied	[22]
8	Granular red mud	Aqueous solution	Cadmium ions	450 rpm, 20 °C, 8–24 h	China	Regeneration was carried out by pumping 0.1 mol/L HCl through the adsorbed column	[23]
9	Modified red mud	Aqueous solution	Arsenate	400 rpm, 24 h, 20 °C	China	pH significantly affects the adsorption and adsorption increases with decrease in pH	[24]
10	$CaCO_3$-dominated red mud	Wastewater	Water-borne Cu, Zn, and Cd	1 h at room temp	China	Water-borne Cu had higher affinity towards red mud than water-borne Zn and Cd	[25]
11	Thermally treated red mud	Wastewater	Nickel ions	24 h at room temp	Bosnia & Herzegovina	Found to be economical adsorbent for nickel (Ni^{2+}) ions removal	[26]
12	Activated red mud	Aqueous solution	Copper	30 °C, 1 h, pH 3&11	Turkey	Activated red mud shows good adsorption of Cu^{2+} as compared to raw red mud	[27]

(continued)

Table 5 (continued)

Sr. no.	Experimental conditions				Country	Remarks	Refs.
	Adsorbent	Waste effluent	Adsorbate	Operating conditions			
13	Ferrous based red mud sludge	Aqueous solution	Arsenic	20 rpm at room temp, 24 h	China	Presence of phosphate can greatly affect the arsenic removal as carbonate had no significant effect	[28]
14	Activated red mud neutralized by CO_2	Aqueous solution	Zinc (II)	180 rpm, 24 h	India	Caustic red mud was neutralized with CO_2 sequestration	[29]
15	Red mud	Underground water	Arsenic (V)	200 rpm, 25 °C	Turkey	Magnetic nano-particles synthesized from red mud and used for adsorption	[30]
16	Granular red mud	Aqueous solution	Cadmium	100 rpm, 30, 40, 50 °C	China	Wet red mud was mixed with cement (2–8%) and granulated	[31]
17	Seawater neutralized red mud and alum water treated sludge	Aqueous solution	As(III), As(V), Se(IV) and Se(VI)	1 h stirring	Australia	Adsorption and desorption was carried out simultaneously	[32]
18	Activated red mud	Aqueous solution	Pb(II)	200 rpm, 1 h	India	Acid neutralized red mud can be used as low-cost adsorbent for adsorption lead from wastewater	[33]
19	Bacteria modified red mud	Aqueous solution	Cadmium ions	pH 4.0, 30 °C	Turkey	Exhibited high potential for the removal of Cd ions from aqueous solution	[34]
20	Iron oxide activated red mud	Aqueous solution	Cadmium	pH 6.0, 27 °C	Turkey	Promising adsorbent for adsorption of cadmium under normal conditions	[35]

(continued)

Table 5 (continued)

Sr. no.	Experimental conditions				Country	Remarks	Refs.
	Adsorbent	Waste effluent	Adsorbate	Operating conditions			
21	Red mud	Aqueous solution	Lead	Batch and column adsorption	Iran	Adsorption capacity was found to be 18.87 mg/g	[36]
22	Acid treated red mud	Aqueous solution	Nickel ions	pH 2–4, batch column	Serbia	Adsorption efficiency decreases with increase in the acid concentration from 0.1 mol/L to 0.25 mol/L	[37]
23	Red mud and nanosized magnetite	Aqueous solution	Arsenic	pH 7	Spain	Red mud was found to be better adsorbent as compared to the nanosized magnetite at pH 7	[38]
24	Seawater neutralised red mud	Aqueous solution	Arsenic	pH 8.5	Denmark	It is not yet able to compete with the other widely used adsorbents but can be used as a pre-treatment method	[39]

Table 6 Adsorption of inorganic anions on treated/amended/neutralized red mud

Sr. no.	Experimental conditions				Country	Remarks	Refs.
	Adsorbent	Waste effluent	Adsorbate	Operating conditions			
1	Pre-treated red mud	Sewage water	Phosphorous	30 min stirring	Japan	Phosphorous level reduced from 20 ppm to 5 ppm	[40]
2	Activated red mud	Aqueous solution	Phosphate	Initial phosphate conc.-50 ppm	–	>50% was removed in 15 min and 72% phosphate removed within 120 min	[41]
3	Activated red mud	Aqueous solution	Phosphate	Adsorption in acetic acid-sodium acetate	India	80–90% phosphate adsorption was observed with initial concentration of 30–100 mg/L	[42]
4	Red mud	Aqueous solution	Phosphate	Crossflow microfiltration technique	Turkey	Phosphate ions act as a coagulant for the red mud particles	[43]
5	Raw and activated red mud	Aqueous solution	Fluoride	pH 1–10 adsorbent conc-1–8.4 g/l	Turkey	About 82% fluoride removal was observed	[44]
6	Raw & activated red mud, fly ash	Aqueous solution	Phosphate	180 rpm, 4 h, 25 °C	China	Red mud is found suitable for phosphate removal from waste effluents from industries such as phosphate fertilizer plants	[45]

(continued)

Table 6 (continued)

Sr. no.	Experimental conditions				Country	Remarks	Refs.
	Adsorbent	Waste effluent	Adsorbate	Operating conditions			
7	Raw and activated red mud	Aqueous solution	Nitrate	Adsorbent conc-1–8 g/L, contact time-5–200 min, initial nitrate conc.-5–250 mg/L, pH- 2–11	Turkey	Activated red mud was found three times more effective than raw red mud	[46]
8	Heat and acid treated red mud	Aqueous solution	Phosphate	180 rpm, 4 h, 25 °C	China	Activated red mud was found effective as compared to the raw red mud	[47]
9	Heat and acid treated red mud	Aqueous solution	Phosphate	70 rpm, 30 & 40 °C	Australia	Acid treated red mud shows greatest adsorption capacity of 0.58 mg P/g at pH 5.5	[48]
10	Granular red mud	Water	Fluoride	400 rpm, 25 °C, 8 h	Turkey	Column adsorption technique was found better than the batch adsorption. Maximum fluoride adsorption was observed at pH 4.7	[49]
11	Granular red mud	Aqueous solution	Phosphate	Desorption using HCl, NaOH and deionised water	China	Phosphate adsorbed red mud can be treated using NaOH for desorption	[50]
12	Granular red mud, bentonite, starch	Aqueous solution	Phosphate	100 rpm, 17, 27, 37 °C, 5 h	China	Sintering temperature affects the phosphate removal	[51]

(continued)

Table 6 (continued)

Sr. no.	Experimental conditions				Country	Remarks	Refs.
	Adsorbent	Waste effluent	Adsorbate	Operating conditions			
13	Granular red mud	Aqueous solution	Orthophosphate &pyrophosphate	100 rpm, 27 °C, 4 h	China	Increase in sintering temperature leads to increase in surface area, but decrease the effective components and enhance the structure stability of adsorbent	[52]
14	Seawater neutralised & heat activated red mud	Aqueous solution	Fluoride	150 rpm, 60 min, 30 °C	India	10% fluoride adsorption has been observed on activated seawater red mud. Acid treatment can increase the adsorption capacity of seawater neutralised red mud	[53]
14	Granular & powdered acid activated red mud, hydroxypropyl methylcellulose and powdered straw	Aqueous solution	Phosphate	100 r/min, 48 h at 20 to 70 °C,	China	Significant results were obtained at 3.0 g/l granular Aan-RM dose, 40 °C adsorption temperature and initial solution 6.0 pH	[54]
15	Acid-activated neutralized red mud	Synthetic solution	Phosphate	100 rpm, 20 min	China	Adsorption capacity increased with increase in initial phosphate concentration and decreased with increase in adsorbent dosage	[55]
16	Modified fly ash& red mud and ferric–alum water treatment residue	Aqueous solution	Phosphorous	30 °C, 48 h	China	Adsorption capacities of synthesised adsorbents is higher than compared to the natural adsorbents	[56]

Table 7 Adsorption of dyes on red mud

Sr. no.	Experimental conditions				Country	Remarks	Refs.
	Adsorbent	Waste effluent	Adsorbate	Operating parameters(Agitation speed time/tempera-ture)			
1	Acid and heat treated red mud	Water	Radiocesium and strontium removal	Ambient temperature	Turkey	Higher can be used as a 'natural barriers' for low-level radioactive wastes and heavy metal-containing products	[57]
2	Washed and dried red mud	Aqueous solution	Congo red	140 rpm, 30 °C	India	Maximum dye removal occurred at the initial pH of 2.0	[58]
3	Washed and dried red mud	Aqueous solution	Acid violet	Adsorption by agitation	India	Maximum dye removal occurred at the initial pH of 4.1	[59]
4	Washed and dried red mud	Aqueous solution	Procion orange	140 rpm, 30 °C	India	Maximum dye removal occurred at the initial pH of 2.0	[60]
5	Treated red mud	Aqueous solution	Rhodamine B, fast green and methylene blue	Batch and column studies	India	About 71.1–94.0% adsorption was achieved by the batch method and 95–97% by column operations	[61]

(continued)

Table 7 (continued)

Sr. no.	Experimental conditions				Country	Remarks	Refs.
	Adsorbent	Waste effluent	Adsorbate	Operating parameters(Agitation speed time/tempera-ture)			
6	Treated red mud and fly ash	Aqueous solution	Methylene blue	100 rpm, 72 h, 30 and 40 °C	Australia	Fly ash generally exhibits higher adsorption capacity than red mud	[62]
7	Microwave activated red mud	Dye wastewater	Scatter blue E-4R	40 °C, pH 3.0	China	Under the optimised conditions, the adsorption rate of scatter blue was found to be 99.07%	[63]
8	MgCl/red mud	Aqueous solution	Colour	150 rpm, 5 min	China	A good quantity of dye removal occurred at pH 12	[64]
9	Chemical and heat-treated red mud	Aqueous solution	Methylene blue	175 rpm, 2 h, 23 °C	Turkey	Chemically treated red mud shows significant adsorption as compared to heat treated red mud	[65]

(continued)

Table 7 (continued)

Sr. no.	Experimental conditions				Country	Remarks	Refs.
	Adsorbent	Waste effluent	Adsorbate	Operating parameters(Agitation speed time/tempera-ture)			
10	Powdered sulfuric acid-treated RM	Dye-contaminated water	Remazol brilliant blue dye	145 rpm, 30 °C	India	About 94% removal of 10 mg/l dye was observed	[66]
11	Activated sintering process red mud	Aqueous solution	Malachite green and crystal violet cationic dye	180 rpm, 180 min	China	Significant adsorption was observed at pH 3.2	[67]
12	Clay and red mud mixtures, Portuguese clay and Fe-impregnated clay	Aqueous solution	Acid orange 7 d	Fenton and photo-Fenton oxidation processes	Portugal	Fenton process shows improvement in the discolouration process (92% within 480 min)	[68]
13	Porous inorganic polymer (IP) monolith ssynthesised using modified red mud	Synthetic wastewater	Methylene Blue (MB)	Constant stirring at 23 ± 2 °C	Belgium	Methylene blue uptake of about 17 mg of MB/g of IP from the solution with an initial MB concentration of 75 mg/L	[69]

Table 8 Adsorption of phenols and its derivatives on red mud

Sr. no.	Experimental conditions				Country	Remarks	Refs.
	Adsorbent	Waste effluent	Adsorbate	Operating parameters (Agitation speed time/temperature)			
1	Treated red mud	Aqueous solution	Phenol, 2-chlorophenol, 4-chlorophenol, and 2,4-dichlorophenol	Agitation and centrifugation	India	About 98% removal of phenols and its derivatives was achieved	[61]
2	Neutralized red mud	Aqueous solution	Phenol	700 rpm, 25 °C, 10 h	Turkey	Higher adsorption rates were observed at low initial phenol concentration in the solution	[70]

2.5 Phenols and Its Derivatives Removal

Phenols and its derivatives such as chlorophenols, nitrophenols are present in the waste effluents of industries such as pharmaceuticals, coking operations, resin manufacturing, petrochemicals, oil refining, paper and pulp industry [10]. Author have studied the removal of chlorophenols (2,4-dichlorophenol and 4-chlorophenol) from wastewater using red mud. Neutralised red mud was used to remove phenol from the aqueous phase in a batch adsorption technique. Table 8 shows concise information on the adsorption of phenols and its derivatives from industrial wastewater using red mud.

2.6 Removal of Other Impurities

Red mud can be utilized for scrubbing of industrial flue gases. A process was developed for manufacturing a gas purifier in which red mud was used as an adsorbent to scrub the gases. Experiments were carried out for neutralizing acidic mine water containing iron, sulphates, and chlorides using red mud. Dairy wastewater was treated with red mud in the context of turbidity, BOD, COD, bacterial count, oil and grease and sludge formation. Detailed study regarding adsorbents, adsorbate, and operating conditions for adsorption of other impurities using red mud has been provided in Table 9.

Table 9 Adsorption of other impurities on red mud

Sr. no.	Experimental conditions				Country	Remarks	Refs.
	Adsorbent	Waste effluent	Adsorbate	Operating parameters			
1	Red mud	Mine water	Iron, sulphates and chlorides	10 g of red mud in 1L of mine water	Germany	pH of mine water raised from 2.5 to 4.8 and Cl^- concentration reduced from 52 to 18 mg/L	[71]
2	Acid treated red mud	Wastewater	–	Treated red mud mixed with waste water	Hungary	There was decrease in COD from 10,000 to 300 mg/L, total suspended solids from 1740 to 175 mg/L, organic suspended materials from 510 to 60 mg/L	[72]
3	Red mud and alum	Dairy wastewater	–	–	India	Significant reduction in turbidity, BOD, COD, oil and grease content and bacterial counts were obtained	[73]
4	Acid treated, seawater neutralized, ammonia treated red mud	Aqueous solution	Ferricyanide	Initial ferricyanide conc – 100 ppm, adsorption time – 120 min, speed – 180 rpm, pH-7, &temp –25 °C	Iran	Maximum adsorption capacity (12.40 mg/g) was observed using activated bauxol using CTAB surfactant	[74]

3 Identified Issues for the Utilization of Red Mud in Pollution Control

Lot of work has been carried out for red mud application in pollution control but no method has been commercialized yet due to some of the technical and environmental issues. These issues have been discussed below.

- Enhancement of sorption efficiency of red mud adsorbents through modification.
- Use of red mud adsorbents for more than one pollutant present in the waste effluent.
- Regeneration of adsorbents made from red mud after its utilization for pollutants removal.
- Proper disposal of contaminated adsorbents.
- Since the quantity consumed by this application is small therefore, industry have not evinced much interest as they are naturally interested in those applications which enable large scale consumption of red mud produced in their industries.

4 Discussion

Red mud/bauxite residue is an alkaline waste generated during the extraction of aluminium from bauxite using Bayer's process. It poses environmental problems due to its high alkalinity. Hence proper disposal and its utilization becomes necessary. It can be utilized in the various application areas. Pollution control is one of the utilization options of the red mud in which it can be treated and used for the adsorption of different toxic elements from flue gases and industrial waste effluents/wastewater. Its chemical composition and physical properties make it suitable for adsorption. Sumitomo is the process patented for the adsorption of SO_2 from exhaust gases showing about 97% of adsorption efficiency. In the removal of metal ions from waste effluents using red mud, adsorption increases with decrease in pH. Activated red mud is found suitable for the phosphate removal from waste effluents from industries such as phosphate fertilizer plants. Phosphate adsorbed red mud can be treated using NaOH for desorption. Acid and heat-treated red mud shows higher adsorption activity for radio-cesium and can be used as 'natural barriers' for low-level radioactive wastes and heavy metal-containing products. Red mud can also be utilized to reduce the COD, BOD, organic solids, total suspended solids from wastewater.

5 Conclusion

Industrial solid and liquid waste management is the major problem faced by the industries all over the world. It is therefore essential to give suitable treatment to these industrial effluents for the removal of toxic elements prior to safe disposal. Red mud is a solid waste generated in alumina industry which can be utilized for the treatment of waste effluents of other industries such as pharmaceuticals, petrochemicals, oil refining, paper and pulp industry, textile industry, sulphuric acid plants, thermal power plants, nitric acid manufacturing, explosive industry, coking operations and resin manufacturing. Treated red mud shows better results but it is not yet able to compete with the other widely used adsorbents and can be used as a pre-treatment method. Therefore, despite lots of research conducted in this area, very few have been commercialized. Subsequently economic feasibility of the process needs to be explored.

References

1. Evans K (2016) The history, challenges and new developments in the management and use of bauxite residue. J Sustain Metall 2(4):316–331. https://doi.org/10.1007/s40831-016-0060-x
2. Xue S et al (2019) Industrial wastes applications for alkalinity regulation in bauxite residue: A comprehensive review. J Cent South Univ 268–288
3. Habuda-Stanic M, ErgovicRavancic M, Flanagan A (2014) A review on adsorption of fluoride from aqueous solution. Materials 7:6317–6366
4. Barakat MA (2011) New trends in removing heavy metals from industrial wastewater. Arab J Chem 4(4):361–377
5. Fu Y, Viraraghavan T (2001) Fungal decolorization of dye wastewaters: a review. Biores Technol 79(3):251–262
6. Robinson T, McMullan G, Marchant R, Nigam P (2001) Remediation of dyes in textile effluent: a critical review on current treatment technologies with a proposed alternative. Biores Technol 77(3):247–255
7. Air pollution. http://www.air-quality.org.uk/04.php. Accessed 10 Dec 2019
8. Mitsubishi Ship Building & Engg Co Ltd (1964) French Patent 1,350,231 24 Jan 1964
9. Yaseen DA, Scholz M (2019) Textile dye wastewater characteristics and constituents of synthetic effluents: a critical review. Int J Environ Sci Te 16(2):1193–1226
10. Villegas L et al (2016) A short review of techniques for phenol removal from wastewater. Current Pollut Rep 2(3):157–167
11. Whitten G et al (1970) J Air Pollut Countr Ass 1:1–4
12. Iida T (1974) Jap Patent 74,84,978, 15 Aug 1974
13. Sato O, Takada Y (1978) Jap Patent 7,81,06,688, 16 Sept 1978
14. Yamada K, Harato T, Shiozaki Y (1979) Ger patent 29,07,600, 06 Sept 1979
15. Fois E, Lallai A, Mura G (2007) Sulphur dioxide adsorption in a bubbling reactor with suspensions of Bayer red mud. Ind Eng Chem Res 46(21):6770–6776
16. Miko S et al (1974) Mizu Shori Gijutsu 8:817–822
17. Fukuda M, Azuma H (1979) Jap Patent 7,91,59,410, 17 Dec 1979
18. Pradhan J, Das SN, Thakur RS (1999) Adsorption of hexavalent chromium from aqueous solution by using activated red mud. J Colloid Interf Sci 217(1):17–141
19. Altundogan HS et al (2000) Arsenic removal from aqueous solutions by adsorption on red mud. Waste Manag 20(8):761–767
20. Genc H et al (2003) Adsorption of arsenate from water using neutralized red mud. J Colloid Interf Sci 264(2):327–334
21. Genc-Fuhrman H et al (2004) Adsorption of arsenic from water using activated neutralized red mud. Environ Sci Technol 38(8):2428–2434
22. Cengeloglu Y et al (2007) Removal of boron from aqueous solution by using neutralized red mud. J Hazard Mater 142(1–2):412–417
23. Zhu C et al (2007) Removal of cadmium from aqueous solutions by adsorption on granular red mud (GRM). Sep Purif Technol 57(1):161–169
24. Zhang S et al (2008) Arsenate removal from aqueous solutions using modified red mud. J Hazard Mater 152(2):486–492
25. Yingqun M et al (2009) Competitive removal of water-borne copper, zinc and cadmium by a $CaCO_3$-dominated red mud. J Hazard Mater 172(2–3):1288–1296
26. Smiljanić S et al (2010) Rinsed and thermally treated red mud sorbents for aqueous Ni^{2+} ions. Chem Eng J 162(1):75–83

27. Nadaroglu H, Kalkan E, Demir N (2010) Removal of copper from aqueous solution using red mud. Desalination 251(1–3):90–95
28. Yiran L et al (2010) Arsenic removal from aqueous solution using ferrous based red mud sludge. J Hazard Mater 177(1–3):131–137
29. Sahu RC, Patel R, Ray BC (2011) Adsorption of Zn(II) on activated red mud: neutralized by CO_2. Desalination 266(1–3):93–97
30. Akin I et al (2012) Arsenic(V) removal from underground water by magnetic nano particles synthesized from waste red mud. J Hazard Mater 235–236:62–68
31. Ju S et al (2012) Removal of cadmium from aqueous solutions using red mud granulated with cement. T Nonferr Metal Soc 22(12):3140–3146
32. Ya-Feng Z, Haynes RJ (2012) Comparison of water treatment sludge and red mud as adsorbents of As and Se in aqueous solution and their capacity for desorption and regeneration. Water Air Soil Poll 223(9):5563–5573
33. Sahu MK et al (2013) Removal of Pb (II) from aqueous solution by acid activated red mud. J Environ Chem Eng 1(4):1315–1324
34. Kalkan E, Nadaroglu H et al (2013) Bacteria-modified red mud for adsorption of cadmium ions from aqueous solutions. Pol J Environ Stud 22(2):417–429
35. Hizal J et al (2013) Heavy metal removal from water by red mud and coal fly ash: an integrated adsorption–solidification/stabilization process. Desalin Water Treat 51:37–39
36. Mobasherpour I, Salahi E, Asjodi A (2014) Research on the batch and fixed- bed column performance of red mud adsorbents for lead removal. Can Chem Trans 2(1):83–96
37. Smičiklas I et al (2014) Effect of acid treatment on red mud properties with implications on Ni (II) sorption and stability. Chem Eng J 242:27–35
38. López-García M et al (2017) New polymeric/inorganic hybrid sorbents based on red mud and nanosized magnetite for large scale applications in As (V) removal. Chem Eng J 311:117–125
39. Fuhrman HG (2018) Arsenic removal from water using seawater-neutralised red mud (Bauxsol). Environment and resources. Technical university, Denmark, Sep 14
40. Sanga S (1974) Jap Patent 74,74,951, 15 Jul 1974
41. Shiao SJ, Akashi K (1977) J Water Pollut Control Fed 49(2):280–285
42. Pradhan J, Das J, Das S, Thakur RS (1998) Adsorption of phosphate from aqueous solution using activated red mud. J Colloid Interf Sci 204(1):169–172
43. Akay G et al (1998) Phosphate removal from water by red mud using crossflow microfiltration. Water Res 32(3):717–726
44. Çengeloğlu Y, Kır E, Ersö M (2002) Removal of fluoride from aqueous solution by using red mud. Sep Purif Technol 28(1):81–86
45. Li Y et al (2006) Phosphate removal from aqueous solutions using raw and activated red mud and fly ash. J Hazard Mater 137(1):374–383
46. Cengeloglu Y, Tor A, Ersoz M, Arslan G (2006) Removal of nitrate from aqueous solution by using red mud. Sep Purif Technol 51(3):374–378
47. Chang-Jun L et al (2007) Adsorption removal of phosphate from aqueous solution by active red mud. Int J Environ Sci 19(10):1166–1170
48. Huang W et al (2008) Phosphate removal from wastewater using red mud. J Hazard Mater 158(1):35–42
49. Tor A et al (2009) Removal of fluoride from water by using granular red mud: batch and column studies. J Hazard Mater 164(1):271–278
50. Zhao Y et al (2010) The regeneration characteristics of various red mud granular adsorbents (RMGA) for phosphate removal using different desorption reagents. J Hazard Mater 182(1–3):309–316
51. Yue Q et al (2010) Research on the characteristics of red mud granular adsorbents (RMGA) for phosphate removal. J Hazard Mater 176(1–3):741–748
52. Zhao Y et al (2012) Influence of sintering temperature on orthophosphate and pyrophosphate removal behaviors of red mud granular adsorbents (RMGA). Colloids Surf A Physicochem Eng Aspects 394:1–7
53. Rai S et al (2013) Seawater neutralised red mud as an adsorbent. Res J Eng Tech 4(2):57–61

54. Ye J et al (2015) Phosphate adsorption onto granular-acid-activated-neutralized red mud: parameter optimization, kinetics, isotherms, and mechanism analysis. Water Air Soil Poll 226:306
55. Ye J (2016) Operational parameter impact and back propagation artificial neural network modeling for phosphate adsorption onto acid-activated neutralized red mud. J Mol Liq 216:35–41
56. Wang Y (2016) Comparison study of phosphorus adsorption on different waste solids: fly ash, red mud and ferric–alum water treatment residues. Int J Environ Sci 50:79–86
57. Apak R et al (1995) Sorptive removal of Cesium-137 and Strontium-90 from water by unconventional sorbents. I. usage of bauxite wastes (Red Muds). J Nucl Sci Technol 32(10)
58. Namasivayam C, Arasi DJSE (1997) Removal of congo red from wastewater by adsorption onto waste red mud. Chemosphere 34(2):401–417
59. Namasivayam C, Yamuna R, Arasi D (2001) Removal of acid violet from wastewater by adsorption on waste red mud. Environ Geol 41(3–4):269–273
60. Namasivayam C, Yamuna RT, Arasi DJSE (2002) Removal of procion orange from wastewater by adsorption on waste red mud. Sep Sci Technol 37(10)
61. Gupta VK, Ali I, Saini VK (2004) Removal of chlorophenols from wastewater using red mud: an aluminum industry waste. Environ Sci Technol 38(14):4012–4018
62. Wang S, Boyjoo Y, Choueib A, Zhu ZH (2005) Removal of dyes from aqueous solution using fly ash and red mud. Water Res 39(1):129–138
63. Gong C, Xia D, Zeng Q (2008) Adsorption removal of scatter blue E- 4R by microwave activated red mud. Ind Water Wastewater
64. Wang Q et al (2009) The color removal of dye wastewater by magnesium chloride/red mud (MRM) from aqueous solution. J Hazard Mater 170(2–3):690–698
65. Çoruh S, Geyikçi F, Ergun ON (2011) Adsorption of basic dye from wastewater using raw and activated red mud. Environ Technol 32(11):1183–1193
66. Ratnamala GM, Shetty VK, Srinikethan G (2012) Removal of remazol brilliant blue dye from dye-contaminated water by adsorption using red mud: equilibrium, kinetic, and thermodynamic studies. Water Air Soil Poll 223(9):6187–6199
67. Zhang L, Zhang H, Guo W, Tian Y (2014) Removal of malachite green and crystal violet cationic dyes from aqueous solution using activated sintering process red mud. Appl Clay Sci 93–94:85–93
68. Hajjaji W, Pullar RC, Labrincha JA, Rocha F (2016) Aqueous acid orange 7 dye removal by clay and red mud mixes. Appl Clay Sci 126:197–206
69. Hertel T, Novais RM, Murillo RM, Labrincha JA, Pontikes Y (2019) Use of modified bauxite residue-based porous inorganic polymer monoliths as adsorbents of methylene blue. J Clen Prod 227:877–889
70. Tor A, Cengeloglu Y, Aydin ME, Ersoz M (2006) Removal of phenol from aqueous phase by using neutralized red mud. J Colloid Interface Sci 300(2):498–503
71. Starosta KH, Glowna K, Ahrens K (1976) Ger(east) patent 1,20,185, 05 June 1976
72. Szirmai E, Babusek S, Horvath G, Makai A, Vadas T, Wenczel M (1989) Hung Patent 47,059, 30 Jan 1989
73. Namasivayam C, Ranganathan K (1992) Res Ind 37(3):165–167
74. Deihimi N, Irannajad M, Rezai B (2018) Characterization studies of red mud modification processes as adsorbent for enhancing ferricyanide removal. J Environ Manage 206:266–275

Aluminum Dross: Value Added Products to Achieve Zero Waste

Upendra Singh

Abstract Aluminium dross is a hazardous by-product of the aluminum smelting industry. The waste Aluminium dross generated is usually dumped which is an environmental threat as it contains many hazardous chemicals. The management of Aluminium dross can be done by conversion to different value-added products. Coagulant application is one such value-added option for Aluminium dross to end dumping. This work investigates the possibility of using aluminum dross for preparation of widely used water treatment coagulants i.e., Alum. The process involves the dissolution of Aluminium dross to prepare alum with sulfuric acid. Metallic aluminum in the waste dross is dissolved into the sulfuric acid solution, and the solution could be used as alum for water treatment chemicals after adjusting the required alumina concentration and pH of the solution. The success of project will open up aluminum dross as new/alternative source of raw material for preparation of Alum/PAC. Also, there is an additional advantage in view of recycling of the waste aluminum dross by reducing the amount of waste disposed to landfill.

1 Introduction

Aluminum dross is a process reject of Aluminium metal production. Dross is classified into different categories based on the metal content such as white dross, black dross, foundry dross, induction furnace dross, primary aluminum dross, etc. The interest in this material is really a small percentage of the total aluminum of the cast house, but it has a substantial economic effect. The value of the recovered aluminum is $2000/ton while value of oxide is 0–$400/ton which emphasis on which material we should maximize in recovering [1]. Aluminum converted to oxides after skimming is an expensive problem.

U. Singh (✉)
Jawaharlal Nehru Aluminium Research Development and Design Centre (JNARDDC), Amravati Road, Wadi, Nagpur 440023, (M.S.), India
e-mail: singhu@jnarddc.gov.in

K. Randive et al. (eds.), *Innovations in Sustainable Mining*, Earth and Environmental Sciences Library, https://doi.org/10.1007/978-3-030-73796-2_3

Therefore, the aim of any handling method used in the cast house is to prevent aluminium from oxidizing any further than when it was removed from furnace. It is known that aluminum starts rapid oxidation in the air when its temperature rises above approximately 780 °C. Generally, the rate of oxidation increases rapidly as the temperature increases.

Typically, 15–25 kg of dross is produced per metric ton of molten Aluminium [2]. Dross can contain 10–80% aluminium depending upon different grades. Besides Aluminium metal dross may contain other chemical compounds e.g., Al_2O_3, AlN, Al_4C_3, MgF_2, SiO_2 and MgO etc. [3]. Aluminium dross is mainly of two main—white dross and black dross. White dross has higher aluminium metal content and is produced from primary and secondary aluminium smelters and re-melt shops, whereas black dross has a lower metal content and is generated during aluminium recycling (secondary industry sector). The dross is processed in rotary furnace to recover the metallic Aluminium, and the remaining is disposed off to landfills in unaltered condition. The non-metallic residues generated from dross smelting processes are often referred to as 'salt cake' and contain 3–5% residual metallic aluminium.

Every year, around 3–4 million tonnes of primary dross and more than one million tonnes of secondary dross is generated worldwide, and >95% of this material is land filled. Cost reduction, increased process efficiency and sustainable practices all make limiting dross formation during the melting process a major objective for cast houses and re-melt shops. Dross handling and processing to recover metal is equally important. With current technology over 65% will be expected as the minimum return of metal from a given dross weight. All this has been possible as a result of continuous developments in the technology to meet the environmental concerns, energy conservation, profitability and community impact for the long-term sustainability of the industry the triple bottom line concept will be essential [4].

Alumina is the key component for refractory preparation used in high-temperature industrial applications; such as boiler, castable, cement, light industry, electric power, military industry, ceramic, glass, and petrochemical industry. World consumption of calcined refractory-grade bauxite is about 1.0 million tons per year and calcined alumina for use in refractory applications is about 500,000 metric tons per year [5]. Therefore, there is market potential for Aluminium dross waste as an alternative source of alumina for refractory materials.

The potential benefit of directing aluminum dross towards extraction of alum and refractory material is important because dross is a great source of aluminum oxides, so it can be an alternative source to primary materials [6]. Moreover, it also reduces the waste disposed to landfills. Due to gradual depletion of natural resources of refractory grade raw material, the concept of converting non-conventional material into usable grade through value addition receiving considerable attention in recent years.

2 History and Future of Dross

Due to the profitable economics of aluminum and the its oxidation products over the past 60 years the Aluminium industry focus has shifted from one that was thrown away to one that we look to complete recycle. The practice of landfilling Aluminium dross has raised concerns for both industry and environment. We have moved forward in dross processing technology development. The first challenge is to keep as much aluminum as possible inside the melting furnace. The next step is to develop technologies to recycle Aluminium from dross to produce added-value products and not a landfill liability.

3 Primary Processes

Followings are the primary processes usually exercised in the soft floor of the Aluminium industry.

- Manual Hand Picking
- Floor Cooling
- Shaker/Vibrators/Stirrers
- Rotary Coolers/Vibrators/Stirrers
- Inert Gas Dross Cooling (IGDC)
- Hot Dross Press
- Mechanical Processing
- Value added product development

4 Future Scope

Recycling and reutilization of industrial waste and by-products are the subjects of great importance today in industry. During the production of aluminum, a huge amount of waste is generated, known as dross, which is usually recycle for recovery of metallic aluminium by metallurgical processes. There is no organized sector to handle/manage aluminium dross except metallic aluminium recovery and rest of residue is disposed off in landfill. The residues contain toxic elements such as nitride, fluoride, carbide and others which are likely to result in leaching of toxic metal ions into ground water causing serious pollution problems and threat to the ecosystem. It is a big challenge for researchers to develop a process for complete utilization of waste aluminium dross to achieve zero waste. In this context, a suitable research work has been developed. The water treatment agent such alum could be produced using chemical leaching and remaining dross as high alumina refractory castables. The process could achieve zero waste concept using waste dross material [7].

5 Technical Assessment of Dross

5.1 Sample Collection & Preparation

About 50 kg of Aluminium dross sample was collected manually from dross yard from Indian Smelter. The whole sample was packed in a plastic bag. The utmost care was taken while transporting the sample to R&D Centre (Fig. 1).

5.2 Screening/Classification

The whole sample was treated as a lot. Dross sample was pulverized and classified in two main parts, lumpy and powder dross. Big size metals were separated by hand picking and kept separately. The rest of the sample was screened using standard sieve of mesh no 5 (i.e., 4 mm). The powder (fine size dross) was screened from medium size particles. The undersize was treated as low grade Aluminium dross as it contains lower metal value and un-economical for recovery of metal by conventional metallurgical processes such as melting etc. (Fig. 2).

- **Physico-chemical characterization of dross**: The most widely used methods of characterization have been utilized in the characterization of Aluminium dross sample, intermediate product and alum.

Fig. 1 Sample collections at plant site

Fig. 2 Classification of sample using sieve

5.3 *Wet Chemical*

The conventional wet chemical method was used for analysis of various elements present in Aluminium dross and quantified. Usually, triple acid digestion or hydrochloric acid digestion was used for sample preparation and analysis.

Instrumental

Inductively coupled plasma spectrometer (ICP-AES, Thermo Elemental, USA), Flame photometer (Systronics, India), Ion Selective Electrode (ISE-ORION, Thermo USA), Scanning electron microscope (SEM- JEOL, Singapore), X-ray diffraction (XRD Panalytical Holland), GDS (LECO, USA) TOC, (Schimadzu, Japan) and other equipment have been exclusively utilized in the proposed work. Elemental characterization was done using acid digestion. Thermal analysis was done with TG/DTA and DSC (NETZSCH). The Aluminium metal analyses were done by dissolving known weight of dross sample in hydrochloric acid. Ion selective electrode was utilized in quantification of fluoride for both Aluminium dross and alum in the present study. Quantification of sodium in dross was carried out by flame photometer. Other test such as PCE, PLC, BD, CCS etc. was done for the characterization/application of residual dross in refractory application.

5.4 *Determination of Major Metal/Metal Oxide*

Metallic Aluminium

Undersize (−5 mesh) was further ground and homogenized for analysis of metallic as well as other elemental impurities. For metal analysis 20 gm Al-Dross (−5 mesh) was taken in 500 ml beaker then moist with 2 ml distilled water.20 ml conc. hydrochloric acid (HCl) was added slowly with constant stirring and kept for 15–20 min at room temperature then 100 ml distilled water was added in it and boiled it on hot plate for ½ an hr. After boiling wait for ½ hr for settling of residue then 40 ml HCl was added further in residue and then 100 ml distilled water was added in it and boiled it on hot plate. Aftercomplete reaction it was cooled at room temperature and filter

Table 1 Digsignation of aluminium dross for different grade

S. no.	Dross quality	Average % recovery by melting
1	Altek press	50
2	A	62
3	B	52
4	Waste-B	20
5	Waste-C	15
6	Waste-D	Not possible

through Whatman filter paper no. 40 in 1000 ml volumetric flask. Washed the beaker 2–3 times with distilled water then residue was washed several times with distilled water and make up the volume up to mark.

Determination of Al^{+3} by Titrimetry

- 5 ml above stock solution was taken in 250 ml beaker, 60 ml 2 N NaOH was added and boiled it on hot plate.
- After boiling cooled it and filter it through Filtrum no. 130 filter paper in 500 ml beaker.
- 25 ml 0.05 N ethylene di amine tetra acetic acid di sodium salt was added in it and neutralized it by 1:1 H_2SO_4 and NH_4OH.
- 30 ml Buffer (ammonium acetate and acetic acid) solution was added then boiled it on hot plate.
- After boiling cooled it and titrate it with 0.05 N zinc acetate using Xylenol orange as indicator (Table 1).

Calculation

- Blank = R ml
- **% Al_2O_3** = {[(Blank−Burette Reading) × 2.55]/weight of sample in mg} × 100

6 Methodology

6.1 Recovery of Metal by Metallurgical Process

12 kg Material above 4 mm or +5 mesh is fed in the induction furnace to extract/recover metallic Aluminium. After attaining the temperature, whole solid metal melt which was subsequently cast into ingots. The melting process also produces some residual dross (approx. 1 kg) of very low grade, which may further recycle in the process to recover remaining metallic Aluminium (Fig. 3).

Induction Furnace

Cast Ingots

Recovered Metal

Fig. 3 Metal recovery processes from Aluminium dross

6.2 Designation of Dross Based on Recoverable Metal

Aluminium dross is designated for different grade based on recoverable metals in some of the primary industry in India as below (Table 2).

Table 2 Chemical analysis of aluminium dross

Constituents	%
Al_2O_3	14.8–15.2
Ca	0.158
Cu	0.0148
Fe	0.0527
K	0.2933
Mg	0.0676
Mn	0.0027
Na	0.0173
Ti	0.0126
Fluoride	<10 ppm

6.3 Treatment of Dross for Recovery of Value-Added Product

This process could be more effective on recovery of the remaining metallic aluminum from the dross to separate smaller size from large size before melting. It is better in consideration of melting efficiency because there is much larger amount of metallic aluminum in the large size dross than the fine size. If dross is crushed, size of oxide phase in the dross becomes smaller and metallic phase becomes larger. The critical size in the classification was less than 4 mm in this study. The dross over this size have been re-melted to recover aluminum and the smaller one (the undersize (<5 mesh) was designated as waste dross and processed for chemical leaching.

6.3.1 Chemical Leaching

Initially 1 kg waste Aluminium dross (undersize 5 mesh) was taken and 2.4 L Sulfuric acid (25%) was added slowly with constant stirring in 5 L reaction vessel at ambient temperature. Stoichiometric equivalence of the 98% sulfuric acid was calculated according to the given chemical Eq. (1). The pH of the whole solution was maintained to be around 3, after leaving the whole reaction mixture at ambient atmosphere for 3–4 h, the mixture was filtered with the help of high-volume filtration system to separate residue from liquor (Fig. 4).

$$2\,Al + 3\,H_2SO_4 \xrightarrow{\text{Leaching}} Al_2(SO_4)_3 + 3H_2 \uparrow \quad (1)$$

Since, whole reaction is exothermic in nature; no additional heating is required but oxide adhered to the metal particle does not allow the further reaction, need little warming the solution to complete the reaction. The aluminum sulfate (alum) solution was obtained through filtration. The residue was washed with little amount of dilute

Fig. 4 Chemical leaching

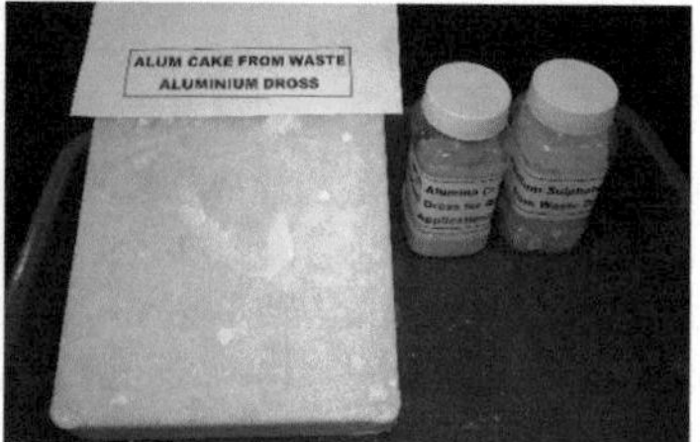

Fig. 5 Alum cake extracted from waste dross

sulfuric acid and finally with water to assure the complete leaching of dissolved salt. The process of chemical leaching was repeated if any relics of Aluminium salts are remained with residue. The filtrate was concentrated to achieve the final specific gravity of liquor to be around 1.38–1.45 by evaporating the excess water. After achieving the required specific gravity, the whole mixture converts into hot cake and gets hardened. The chemical analysis of this cake was done as per standard IS methods which reveals that chemically it was $Al_2\ (SO_4)_3 \times H_2O$ with some impurities such as iron also verified by thermogravimetric characterization.

The chemical equivalence of acid with one equivalent of metallic aluminum and water was considered during preparation for balancing the reacting solution. The experiments were conducted in different sets (1, 2 & 3 kg) to ensure the viability of the process (Fig. 5).

6.3.2 Characterization of Alum

Chemical analysis of alum cake for various elements was done by wet chemical and ICP-AES method. The alumina content in cake was found to be close to the value of A-grade alum commercially available in the market. Also, the soluble iron content is obtained to be very low indicate the purity of product (Table 3).

Table 3 Chemical analysis of Alum extracted from dross

S. no.	Elements %	(−5 Mesh)
1	Al (Metal)	21.4
2	Ca	0.74
3	Cu	0.02
4	Fe	0.38
5	K	0.43
6	Mg	0.22
7	Mn	0.004
8	Na	0.30

7 Application of Alum

Water is a precious commodity. Much of the earth's water is from sea. Fresh water that does not contain large amounts of dissolved salts is around 2.5% of the water and two-third of that is frozen in permafrost.

In total only 0.01% of the total water of the planet is available for use. Clean and safe water for drinking is a basic human necessity. Unfortunately, in the developing world, more than one out of six individuals still lack reliable access to this valuable essential. India accounts for 2.45% of world's land area and 4% of water resources of the world but comprises 16% of the global population. With the current population growth-rate (1.9% per year), the population is expected to cross the 1.5 billion mark by 2050 [8]. The rapid growth of population has exceeded the potable water requirement, which requires exploration of raw water supplies, development of treatment methods and distribution networks.

Disposal of the backwash water and sludge from water treatment plants is a major environmental issue. Therefore, it is essential to optimize chemical dosing and filter runs in order to minimize rejection from water treatment plants. Alum is being added as coagulant in almost all water treatment plants in Aluminium industries and also in Municipal Corporation to meet the requirement. The trend of urbanization and industrialization in India is exerting stress on civic authorities to provide their citizens and employees with basic requirements, such as safe drinking water.

In this direction an effort of utilizing waste for converting a value-added product such alum has been initiated. This could be a new resource of generating water aid chemicals for the purpose of raw water treatment and could be used effectively for the commercial production of alum.

Solids are present in water in three primary forms: suspended particles, colloids and dissolved molecules. Suspended particles vary in size from very large particles down to particles with a typical dimension of 10 μm, such as sand, vegetable matter and silts, Colloids are very fine particles, usually ranging from 10 nm to 10 μm.

Dissolved molecules are present as molecules or as ions. Conventional physical treatment like sedimentation and filtration essentially eliminates the suspended particles. Thus, the removal of suspended particles is the main objective in conventional water treatment. The treatment of raw water with commercial alum was compared with performance of developed alum in the study. Flocculation time is usually 10–30 min; however, the optimum flocculation time will vary depending on the raw water quality and downstream clarification technique. Sedimentation is the process of allowing the floc formed during flocculation to settle out and separate from the clarified water. Typical retention time for settling is 6–8 h (Fig. 6).

It was observed that the performance of alum on turbidity reduction was same or even better in case of lab developed alum compared to commercial alum. Primary turbidity in the original sample was 14 NTU but after a period of 8 h treatment the turbidity was reduced below <1 NTU in lab developed alum

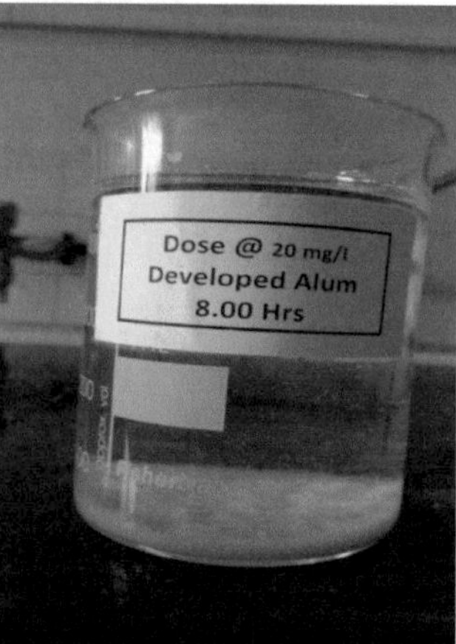

Fig. 6 Coagulation, flocculation and sedimentation process

8 Conclusions

Various important aspects of recovery of metal value, preparation of alum and use in castable refractory from Aluminium dross have been presented in current chapter. The following conclusion may be drawn from the results obtained in the project.

- Aluminium metal content was found to be high, around 42% (undersize sample >20%) and quantified. Alum from undersize Aluminium dross has been prepared and characterized. The characteristic of lab prepared alum is found to have close resemblance with the alum commercially available in the market.
- The alum produced from dross would be much cheaper (approx half) as compared with the alum commercially available in the market, the study has a scope of internal generation of alum from waste grade dross.
- Production of alum from waste Aluminium dross will minimize the costs of the company's procurement of alum.
- The characteristics of alum prepared from dross shows encouraging results. The alum was also used for raw water treatment and found to have more efficient compared the commercial alum.
- The water quality test indicates that water treated with lab developed alum was fit for drinking purposes.
- For the preparation of alum and high alumina refractory castables, aluminium dross could be a new source that could achieve zero waste concepts through total waste utilisation.
- Non-metallic residues (>90% alumina) obtained during the treatment of Aluminium dross could be used as a source of Aluminium oxide in castable
- The end products of the dross management process (Aluminium ingots, powder metallurgy products) are considered as high value.

- Less waste is disposed off to landfill will also help in maintaining the eco-friendly environment.
- Zero waste concepts could be achieved after commercialization of the process.

References

1. Roth D (2015) History and future of dross processing. Light Metal 1005–1009
2. Sultana UK, Gulshan F, Gafur MA, Kurny ASW (2013) Kinetics of recovery of alumina from aluminium casting waste through fusion with sodium hydroxide. Am J Mater Eng Technol 1(3):30–34
3. Ibarra Castro MN, Almanza Robles JM, Cortés Hernández DA et al (2009) Development of mullite/zirconia composites from a mixture of aluminum dross and zircon. Ceram Int 35(2):921–924
4. Draft Strategy on Aluminium Resource Efficiency (2018) Jawaharlal Nehru aluminium research development and design centre, Nagpur
5. Yoshimura HN, Abreu AP, Molisani AL et al (2008) Evaluation of aluminum dross waste as raw material for refractories. Ceram Int 34(3):581–591
6. Mukhopadhyay J, Ramana YV, Singh U (2005) Extraction of value-added products from Aluminium dross material to achieve zero waste. Light Meter 1209
7. Singh U, Ansari MS, Puttewar SP, Agnihotri A (2016) Studies on process for conversion of waste Aluminium dross into value added products. Russian J Non Ferrous Metals 57(4):296–300
8. Status of water treatment plants in India (2008) Central Pollution Control Board

Value Addition of Alumino-Silicates: Consolidation of Mining Rejects and Industrial Slag by Geo-Polymerization

P. A. Mohamed Najar, Amrita Karn, Vishakha Sakhare, Mukesh Jitsingh Chaddha, and Anupam Agnihotri

Abstract Industrial rejects are considered secondary resources for several applications involving recovery of mineral values, preparation of low-cost precursors for onward processing as well as raw materials for product development. The scope of utilization of industrial rejects broadly considers material detoxification, process efficiency, overall economics, energy use, impact on environment and the extent of byproduct generation. Aluminium, steel, power and biomass-based industries seemingly generate solid rejects at various stages of production process and remained unusable leftover material. In recent past, selectivity of material input based on synergistic utilization of industrial rejects of different origin has been found useful for converting rejects in to quality product. Choice of making mix designs based on chemical and mineralogical balance among the reject material enabled reduced consumption of commercial ingredients. Also, the choice of selectivity in raw material input provide flexible control in the regulation of physical properties of geo-polymer products such as crushing strength, surface finish, porosity, water absorption, efflorescence and leaching properties. Products of hard, light-weight and multi layered type comprising single, double and tertiary component mix design used for developing geo-polymer based products which are useful for interior and exterior construction needs such as flooring, walling, decoration, paving etc. It is observed, geo-polymer made from the combination of alumino-siliceous industry rejects are a cleaner source of raw material input for low-carbon infrastructure generation which holds ample scope for consuming large volume of underutilized industrial leftover.

Keywords Industrial rejects · Material synergy · Geo-polymerization · Low-carbon infrastructure · Product development

P. A. Mohamed Najar (✉) · A. Karn · V. Sakhare · M. J. Chaddha · A. Agnihotri
Jawaharlal Nehru Aluminium Research Development and Design Centre, Amravati Road, Nagpur 440 023, India
e-mail: najar@jnarddc.gov.in

M. J. Chaddha
e-mail: mjchaddha@jnarddc.gov.in

K. Randive et al. (eds.), *Innovations in Sustainable Mining*, Earth and Environmental Sciences Library, https://doi.org/10.1007/978-3-030-73796-2_4

1 Introduction

Chemistry of aluminosilicate focused on minerals composed of aluminium, silicon and oxygen and corresponding counteractions. They form a major component of kaolin and other clay minerals of hybrid materials [1–3]. The solubility of aluminium, silicon and aluminosilicates in aqueous alkaline solution is of interest in several industrial applications. Aluminosilicates will undoubtedly continue to play the central role in many areas of basic and applied research, such as chemical catalysis, classical and advanced materials production, theoretical chemistry and geochemistry. The class of aluminosilicates is available at many secondary sources; mostly the mining rejects are getting more attention in view of environmental concern relevant to the protection of natural resources. Metallurgical industry is known to generate an array of solid rejects starting from mining activities to finished products. Both aluminium and steel industries are prominent source of alumino-silicate based solid rejects, which possess typical chemical and mineralogical composition that suits well for alkali activation. Different types of ashes, including biomass ash like bagasse ashes, rice husk ash, corncob ash etc. are also of significant source of aluminosilicates.

The major mining rejects of aluminium industry involve alumino-siliceous resources such as high silica and low alumina fractions of bauxite (low grade bauxite fractions), khondalites, lateritic overburden etc. Bauxite residue known as red mud is the largest process reject generated by aluminium industry in the refining of bauxite by Bayer process and its chemical composition is adequately high to utilize red mud as a mineral rich resource for making construction and special application in bricks, polymerized wood substitute, soil amendment as well as for direct commercial exploitations such as cement and steel industries [4]. The solid rejects generated in steel industry is broadly divided into two categories i.e., ferruginous and non-ferruginous rejects. Ferruginous rejects are iron bearing and generated from steel making viz., mill scale, flue dust, sludge from Gas cleaning plants of Blast Furnaces and Steel Melting Shops, Blast furnace slag and SMS slag. Non-ferruginous rejects are lime fines, broken refractory bricks, broken fire clay bricks, acetylene plant sludge etc.

Many researchers have evaluated the use of different ash as constituents in mortar/concrete, including blast furnace slag; olive-stone biomass ash and rice husk; red mud; palm oil fuel ash; fly ash [5–7]. The perception of energy efficient building materials introduced the concept of a chemical process generally known as "geo-polymerization" for development of green building materials from a wide variety of industrial rejects. The chemical reaction between silica, alumina, calcium, magnesium etc. of resource material under alkaline condition forms hydrous structures of rock forming silicate minerals. Production of low-cost building materials by non-fired process provides the impetus in low cost construction and building applications in rural and urban sectors. It is also known that, certain natural and reject materials produced by the alkali activation effectively immobilize heavy metals and other toxic components. Toxic elements that are present in hazardous reject materials can be captured in 3-dimensional network structures [8, 9].

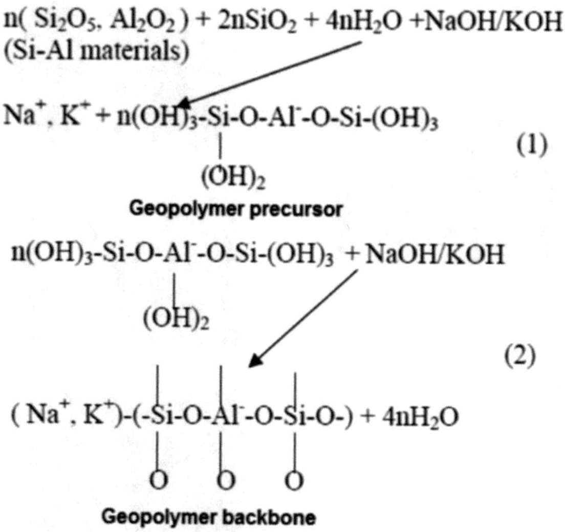

Fig. 1 Chemical reaction during geo-polymerization

$$n(Si_2O_5, Al_2O_2) + 2nSiO_2 + 4nH_2O + NaOH/KOH$$

(Si-Al materials)

$$Na^+, K^+ + n(OH)_3\text{-Si-O-Al}^-\text{-O-Si-(OH)}_3 \quad (1)$$

$(OH)_2$

Geopolymer precursor

$$n(OH)_3\text{-Si-O-Al}^-\text{-O-Si-(OH)}_3 + NaOH/KOH$$

$(OH)_2$

$$(Na^+, K^+)\text{-(-Si-O-Al}^-\text{-O-Si-O-)} + 4nH_2O \quad (2)$$

O O O

Geopolymer backbone

The chemical composition of the geo-polymer material is like natural zeolitic materials, but the microstructure is amorphous instead of crystalline [10, 11]. The polymerization process involves substantially fast chemical reaction under alkaline condition on Si–Al minerals that results in a three-dimensional polymeric chain and ring structure consisting of Si–O–Al–O bonds. Unlike ordinary Portland/pozzolanic cements, geo-polymers do not form Calcium-Silicate-Hydrates (C–S–H) for matrix formation and strength. It utilizes the poly-condensation of silica and alumina and a high alkali content to attain structural strength. Consequently, geo-polymers are occasionally referred “alkali activated alumino silicate binders” [12].

The chemical reactions in the formation of geo-polymer in alkaline conditions are shown in Fig. 1.

These three steps can occur almost simultaneously and may also overlap. Reaction (2) indicates that during polymerization, water (H_2O) is eliminated. [On the other hand, during hydration of ordinary Portland Cement (OPC), water is consumed].

The following steps involved during the geo-polymerization reaction (Fig. 2).

- Si and Al atoms in raw materials dissolve by the action of hydroxide ions
- Precursor ions may be converted into monomers
- Polycondensation of monomers into polymeric structures.

Building materials based on green technology literally reduces carbon dioxide (CO_2) emission in the atmosphere. Accordingly, development of green building materials based on alkali activated geo-polymers received increased R&D attention in recent years and corresponding studies revealed that geo-polymers have excellent mechanical properties such as high early strength and low shrinkage. The feasibility of converting industrial rejects into geo-polymer is analyzed based on criteria of compressive strength, density, embodied energy, CO_2 emissions for

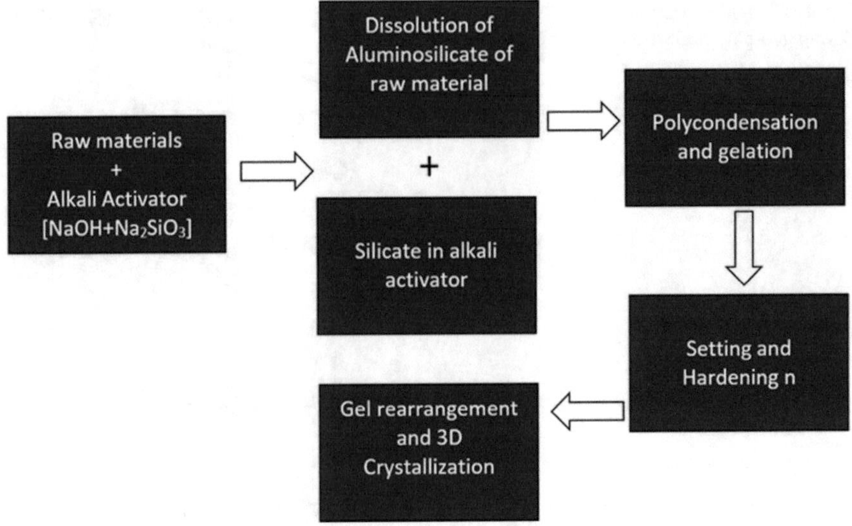

Fig. 2 Process mechanism in geo-polymer formation

different molarity of sodium hydroxide and curing condition. The developed geo-polymer products have similar mechanical as well as physical properties compared to conventional building products. This research work investigated the scope of synergy in the potential utilization of variety of industrial rejects as raw materials to produce geo-polymers that can be effectively used as a masonry product.

2 Experimental

2.1 Materials and Methods

The raw materials used for geo-polymer synthesis include industrial rejects such as red mud, lateritic overburden, partially lateritic khondalite and kaolinitic khondalite etc. that were collected from different mining sites and alumina refineries. The discards of other industrial origin such as fly ash, rice husk ash, etc. were obtained from local industrial units. Commercially available sodium hydroxide and sodium silicate (Qualigens, India) were used for preparing activator medium. All other chemicals and reagents used were of analytical grade.

Sample preparation trials, gravimetric and titrimetric trials were carried out at the chemical laboratory equipped with all standard analytical tools. Mettler Toledo (Switzerland) pH meter was used for all pH measurements. The photometric measurements were carried out on Dual Wavelength UV Visible spectrophotometer (Thermo Scientific Evolution 201 Model 840–210,800). Trace analysis was carried out by inductively coupled plasma spectrometer (IRIS Intrepid II XDL, Thermo). X-ray diffraction (XRD) measurements were performed on both untreated and chemically treated red mud samples using a Philips X-ray diffractometer. Scanning Electron Micro Probe Analyzer-JEOL, JAPAN JXA-840 was used for studying surface morphology.

2.2 Sample Preparation and Characterization

The reject collected from steel industry was dried at room temperature to remove excess moisture, homogenized and sieved. The samples were passed through different meshes to collect the fractions 100, 150 and 200 for the studies and chemically characterized for verifying change in concentration of major constituents such as Fe_2O_3 and CaO with change in size fractions. Since no reasonable variations observed, 100 mesh fractions were selected for further studies.

The samples were crushed, pulverized and homogenized before sample preparation. Powdered samples were sieved (passed a #100 mesh) and were subjected for chemical, mineralogical (X-Ray Diffraction; XRD) and morphological (Scanning Electron Microscopy; SEM) analysis. Based on the chemical composition and mineralogy of raw materials, the combination matrix for raw materials was decided.

The raw material components (selected from prepared stocks of specific mesh size) for making mix design were weighed in electronic balance and mixed in appropriate ratios (based on the respective chemical composition) and thoroughly homogenized in a high-speed mixer to form the dry mixture composition. Subsequently, the alkali activator solution of appropriate molar concentration (Mixture of NaOH & Na_2SiO_3 in specific ratios) was mixed with the dry mix to make flowable wet slurry to pour into the molds.

2.3 Optimization of Solid—Liquid Ratio

Optimized solid -liquid ratio could not be established in the preliminary experimental trials because of the change of components and their mix ratios. The solid liquid ratio was optimized later when the mix design was found useful for preparation of geo-polymer product suitable for building applications. The optimized solid-liquid ratio for specific mix design and product has been described wherever it is specifically used.

The formation of geo-polymer based on synergistic combination of materials has been achieved at ambient conditions in view of better cost impact. The geo-polymer products developed from different mix designs were cured at room temperature and its properties depends up on the nature of materials, the curing period extend from 7 to 36 days. Mild heating is recommended depend on the climatic conditions, especially in rainy and winter season. The molds containing alkaline slurry require mild heat treatment in range 50–70° for 3–4 h to settle the excess moisture prior to geopolymer gel formation followed by setting and curing.

2.4 Alkali Activator

The alkaline activator used for geo-polymer synthesis was a mixture of NaOH, sodium silicate solution and water. The aqueous solutions of sodium silicate (Na_2SiO_3) and sodium hydroxide (NaOH) in required molar concentrations were prepared and mixed in the ration 1:1 v/v. Several possible geo-polymer compositions from different combinations of raw materials were designed to produce viable and workable geo-polymers. The solid–liquid ratio regarding raw material and alkali activator varied with respect to the change in components in the raw material mix. Consequently, different volumes of activator mix were used for making geo-polymers.

2.5 Geo-Polymer Synthesis

Specimen samples of geo-polymer were prepared using combinations of different raw materials (mix design) (Table 1). The mix design was prepared based on the silica and alumina contents. Single or double components dry mix was prepared in specific ratios. Subsequently the dry mix was mixed with alkali activator solution (mixture of sodium hydroxide and sodium silicate) to form wet slurry to pour in to rectangular flat molds (7 cm × 7 cm × 7 cm). The molds with wet slurry were kept

Table 1 Composition of geo-polymer mix

Abbreviation	Constituents raw materials	Mix ratio (w/w)	Concentration & Ratio of activator (mole)
Geo-1	Laterite + Fly Ash + RHA	50:25:25	10:6.0 [1:1]
Geo-2	Laterite + Fly Ash + RHA	50:25:25	08:2.5 [1:1]
Geo-3	Laterite + Fly Ash + RHA	40:40:20	10:2.5 [1:1]
Geo-4	Laterite + Fly Ash + RHA	50:25:25	10:2.5 [1:1]
Geo-5	Laterite + Fly Ash	50:50:00	10:6.0 [1:1]
Geo-6	Laterite + Fly Ash + Process Slag	30:30:40	10:6.0 [1:1]
Geo-7	RM + Fly Ash + RHA	30:30:40	10:6.0 [1:2]
Geo-8	RM + Fly Ash + RHA	30:30:40	10:6.0 [1:1]
Geo-9	RM + Fly Ash + RHA	50:25:25	12:10 [1:1]

RM: Red Mud; RHA: Rice Husk Ash

uninterrupted for three days at room temperature and subsequently the specimens were demolded to remain at room temperature for 28 days (ambient) for curing. The specimen samples obtained were observed for crack, strength and deformation. Crushing strength, water absorption and efflorescence of the sample determined as per Indian standards (IS 1077–1992) [13].

2.6 *Preliminary Test for Geo-Polymer Formation*

The formation of geo-polymer after curing was confirmed by a simple physical test. The hard blocks were cut into cubical pieces in 5 × 5 × 5 cm and dipped in water for 10–15 min. The appearance of disintegrated blocks under water indicated non-formation of geopolymer structure in the material combination and the mix design is identified incompatible for geopolymer preparation.

2.7 *Assessment of Chemical Properties*

Geo-polymer brick sample of varying chemical composition were crushed, and the broken pieces were powdered to get samples of 100–200 μ. The sample powder was analyzed by wet analytical procedure like that followed for individual raw materials. The analytical report obtained for the samples prepared at different experimental conditions were assessed for verifying heavy metal contents. With respect to mix ratio of components in the mix design the overall composition product also changed. Accordingly, it was observed that, the excess percentage of certain constitutes are found diluted in the process of preparing multi components mix designs. Variations in the chemical composition of raw materials within the mix design finds good applications for creating better synergy among the raw materials. Further, the synergistic combination of raw materials is more adaptable for using the same mix design for preparing different product range such as light weight and porous structure with good mechanical stability.

2.8 *Tests on the Developed Product*

The developed geo-polymer was tested for compressive strength, water absorption and efflorescence as per IS-3495 (Part-1):1976 [14]. The specimen cubes were tested under Universal Testing Machine (U.T.M.) having load limit of 100 KN at 28 days. Minimum average compressive strength of brick shall not be less than 7.5 N/mm^2 when tested as per IS-3495 (Part-1):1976 [14]. Average of the four specimens was considered for the analysis.

The density was calculated by heating the samples at 60 °C for 24 h and weight of the cooled samples was taken. Average of three cubes was recorded. Water absorption can be defined as percentage of water absorbs in terms of weight of the specimen. The bricks when tested in accordance with the procedure refer to IS: 3495 (Part-2):1976 [14]. Accordingly, after immersion in cold water for 24 h, water absorption should not be more than 20%. Efflorescence test was conducted for determining the presence of excess soda in geo-polymer specimens prepared with specific mix designs. The bricks when tested in accordance with the procedure of IS: 3495 (Part-3):1976 [14].

3 Results and Discussion

The present study explored the possibility of utilizing the major solid reject originated from aluminium industry in combination with rejects of steel industry and locally generated biomass ash. The analytical data of some of the selected raw material obtained from classical wet analysis is summarized in Table 2.

The preliminary efforts on geopolymerization identified the synergy among the materials with respect to the chemical composition of components such as alumina, silica, and calcium. Further, the role of mineralogy in the formation of geo-polymer matrix was also realized. Accordingly, the total mass of freely available silica, alumina or calcium in the raw materials combinations for alkali activation largely influence the quality of geopolymer products. The good synergistic impact among the raw materials supported the development of low-cost building products. Ten different combinations of materials containing varying concentration ranges of

Table 2 Average chemical composition of raw materials

	Composition %						
Sample	Al_2O_3	Fe_2O_3	SiO_2	TiO_2	LOI	Na_2O	CaO
L	40.09	32.34	6.75	1.87	18.95	0.11	0.98
PLK	35.44	17.96	27.36	02.24	15.65	0.08	0.01
KK	53.00	02.00	22.00	1.00	22.00	0.08	0.09
SL	28.11	05.00	34.89	0.01	09.00	0.21	0.67
RM	17.34	52.22	7.54	8.32	12.11	4.50	1.75
FA	30.56	05.39	59.00	2.11	00.30	0.24	0.33
RHA	1.010	01.10	82.32	0.03	04.00	0.20	0.69
BA	12.20	05.67	70.21	Nil	00.80	0.20	5.96
GBFS	04.08	55.20	07.85	0.53	25.38	0.14	2.71

Range (±) 2; Laterite, RM: Red mud; FA: Fly Ash; RHA: Rice Husk Ash; GBFS: Granulated Blast Furnace Slag; BA: Bagasse ash ; PLK: Partially Lateritised khondalite; KK: Lateritic khondalite; SL: Saprolite

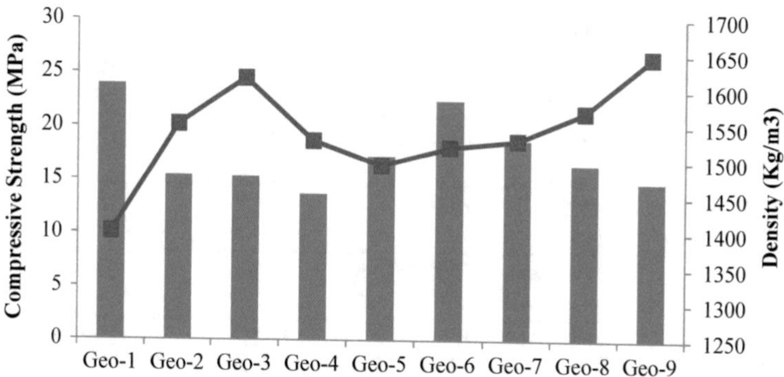

Fig. 3 Compressive strength and density variation of geo-polymer specimens

aluminium, silicon and other constituents were examined to study the scope of geo-polymerization and the impact of activator medium. The impact of alumina and silica concentration in the combination matrix and the silica-soda ratio in alkali activator were found critical in the geo-polymerization process and crushing strength of products.

The compressive strength and density of different mix designs is given in Fig. 3. Minimum average compressive strength of brick shall not be less than 3.5 N/mm^2 when tested as per IS-1077: 1992 [13]. It has been observed that the fineness and density of the material plays a vital role to create dense composition after geo-polymerization. The maximum compressive strength obtained was 23.94 MPa (Geo-1) with materials from aluminium industry origin. All the developed geopolymer combinations are having compressive strength much higher than minimum set by the standards [13]. With the change in activator concentration (Geo-1 and Geo-4), compressive strength increases by 42% while density decrease by 9.2%. The minimum obtained compressive strength is 13.68 MPa (Geo-4) which can be classified under 12.5 class as per IS-1077:1992 whereas highest class (Geo-1) designation which can be achieved using the given mix composition is class 20. Similarly, strength parameter can be achieved using appropriate combination of Process slag (Geo-6) and RM (Geo-7).

The blocks prepared from mix designs showing water absorption less than 20% are shown in Table 3 which represents material combinations of aluminium and steel industries with or without biomass ash as well as fly ash or the combination of fly ash with rice husk ash.

Efflorescence test was conducted for determining the presence of excess soda in geo-polymer specimens prepared with specific mix designs. The bricks when tested in accordance with the procedure of IS-3495 (Part3):1976 [14], the efflorescence rating was not more than moderate for mix designs prepared with aluminium industry rejects. Similarly, no efflorescence rating reported for the mix designs containing rejects from steel industry origin. However, in both aluminium and steel industry rejects mixed with sludge and calcium-based rejects show no efflorescence probably

Table 3 Compositions having less than 20% water absorption

No.	Composition	Mix ratio (w/w)	Con. & ratio of activator (mole)	Comp. strength	Water absorption (%)
1	Laterite + FA + RHA	50:25:25	10:6 [1:1]	23.94	15.00
2	Laterite + FA + Process slag	30:30:40	10:6 [1:1]	22.43	13.96
3	RM + FA + RHA	30:30:40	10:6 [1:2]	18.74	13.64
4	Laterite + FA	050:050	10:6 [1:1]	17.22	10.53

Aluminium industrial reject; RM: Red mud; RHA: Rice Husk Ash

due the early displacement of sodium by calcium leading faster geo-polymer gel formation with alumina-silica in the raw material mix.

Since the efflorescence showing moderate values; appropriate remedial actions need to be taken such as addition of alumina, mild heat treatment, chemical addition and hydrothermal treatment and washing etc. Washing of excess soda in washing chamber is found the cheapest way to remove and recycle the excess alkali. Compared to aluminium industry rejects, steel industry rejects performed better in view of quick setting and low efflorescence formation. Combination of material consisting both aluminium and steel industry origin found to be a better option for restricting efflorescence characteristic of aluminium industry rejects such as red mud and over burden laterite in the geo-polymer mix design. The results of efflorescence tests are shown in Table 4.

Table 4 Efflorescence tests for geo-polymer bricks

No.	Composition	Mix Design	Molar conc. & Ratio of NaOH & Na_2SiO_3	Compressive strength (MPa)	Efflorescence
1	Laterite + FA + RHA	50:25:25	10:6[1:1]	23.94	Moderate
2	Laterite + FA + PS	30:30:40	10:6[1:1]	22.43	Moderate
3	RM + FA + RHA	30:30:40	10:6 [1:2]	18.74	Moderate
4	Laterite + FA	50:50:00	10:6[1:1]	17.22	Moderate
5	RM + FA + RHA	30:30:40	10:6[1:1]	16.43	Moderate
6	Laterite + FA + RHA	50:25:25	8:2.5[1:1]	15.4	Moderate
7	Laterite + FA + RHA	40:40:20	10:2.5[1:1]	15.28	Moderate

PS: Processed blast furnace slag; FA: Fly ash; RM: Red mud; GBFS: Granulated blast furnace slag

Fig. 4 Aluminosilicate (industry rejects) based geo-polymer products

3.1 Value Addition

Value addition of the geo-polymer product was explored by incorporating additional features in the physical properties and aesthetic features. The physical modifications provide improved adaptability geo-polymer as low carbon-low energy building material. The added features would be useful for commercializing geo-polymer products. Accordingly, different type of products such as light weight, hard and stabilized, colored, heat treated, hard as well as light weight foam type, layered (sandwich type) of single and multi-component mix designs, fiber reinforced etc. were derived. These products possess typical features such porous structure, air cavities, low thermal conductivity, better energy absorption, improved tensile strength, light weight and easy workability. Some of the geo-polymer products with additional features are shown in Fig. 4.

3.2 Preparation of Light Weight Foamed Geo-Polymer

Underutilized rejects such as red mud, laterite, fly ash etc., are best suitable raw materials for the synthesis of geo-polymer foams. For the foaming of inorganic polymers reaction between the aluminium metal powder and alkaline activator proceeds quickly, during the reaction releases hydrogen [15, 16]. The wet raw mix was prepared and 1–5% aluminium metal powder was added and homogenized at ambient condition. The wet slurry was transferred in to the molds and allowed to expand for about

Fig. 5 Light weight foamed geo-polymer (Fly ash + GBFS)

1–2 h, followed by curing like normal geo-polymer at atmospheric condition. Light weight foamed geo-polymer in layered and non-layered forms prepared with FA and GBFS combination are shown in Fig. 5.

3.3 Preparation of Layered Geo-Polymer

Two type of layered (sandwich type) geo-polymer products were generated from mix designs of same composition as well as different compositions. In addition, one of the mix designs was made porous for incorporating porous layers in between two hard layers of geo-polymer. The preparation steps involved applying layers of mix designs one over the other in appropriate intervals to avoid deformation due to bonding between different layers.

3.4 Fiber Reinforced Geo-Polymer

Reinforcing with fiber is one of the best options for improving structural stability of geo-polymer blocks. It is very useful when the product is on load bearing applications including construction of road laying. Glass, steel, synthetic and natural fibers are useful for reinforcing. However, it was observed, reinforcing geo-polymer blocks reduces the workability probably due to viscous nature of geo-polymer matrix and uneven distribution of fiber content in the mix.

4 Conclusions

The products developed are found useful as a masonry product in the construction industry. The pre-designed composition (e.g., Si/Al and Na/Si ratios) of the specimens controls the quantity of each raw material or chemical used in each synthesis. The increase in sodium hydroxide molarity has shown significant effect on all the considered parameters. Also, based on the preliminary observations, we confirm the feasibility of converting the industrial rejects in to geo-polymer with calculated addition of silicate-based biomass ash a well as other slag of various origins. The compressive strength which can be achieved using optimum mix of laterite, fly ash and rice husk ash is 23 MPa. A systematic and scientific study further recommend on the topic to optimize all experimental conditions to improve product quality for attaining high strength, light weight and thermal properties for improved value addition and market. It is also envisioned, the typical characteristics of geo-polymer products find its applications other than building materials, especially surface coating, fire retardants and energy storing and regenerating substrate for advanced applications.

Acknowledgements The authors thank the Director, JNARDDC for permission for publishing the work. Science & Technology Wing, Ministry of Mines, Government of India and M/s. Swarnalatha Holdings, Raipur is acknowledged for financial support to the project on *"Synergistic utilization of aluminium industrial wastes for development of geo-polymeric building materials"*.

References

1. Jutzi P, Schubert U (ed) (2003) Silicon chemistry from the atom to extended systems. Wiley-VCH, Weinheim
2. Corriu RJP, Jutzi P (ed) (1996) Tailor-made silicon-oxygen compounds. Vieweg, Braunschweig
3. Favero G, Jobstraibizer, P (1996) Coord Chem Rev 149:367
4. Thakur RS, Das SN (1994) International series on environment−red mud analysis and utilization. Wiley Eastern Limited, New Delhi
5. Font A, Soriano L, Pinheriiro M, Tashima M, Monzo J, Borrachero M, Paya J (2020) Design and properties of 100% waste-based ternary alkali-activated mortars: blast furnace slag, olive-stone biomass ash and rice husk ash. J Cleaner Prod 243:118568
6. Sukmak P, Sukmak G, Horpibulsuk S, Setkit M, Kassawat S, Arulrajah A (2017) Palm oil fuel ash-soft soil geopolymer for subgrade applications: strength and microstructural evaluation. Road Mater Pavement Design 20(1):110–131
7. Guades E (2016) Experimental investigation of the compressive and tensile strengths of geopolymer mortar: the effect of sand/fly ash (S/FA) ratio. Constr Build Mater 127:484–493
8. Gout B, Pan D, Liu B, Volinsky AA, Fincan M, Du J, Zhang S (2017) Immobilization mechanism of Pb in fly ash-based geopolymer. Constr Build Mater 134:123–130
9. Huyen VuT, Gowripalan N (2018) Mechanisms of heavy metal immobilization using geopolymerization techniques–a review. J Adv Concr Technol 16:124–135
10. Palomo A, Grutzeck MW, Blanco MT (1999) Alkali-activated fly ashes a cement for the future. Cem Concr Res 29(8):1323–1329
11. Xu H, Deventer JSJV (2000) The geo-polymerisation of alumino-silicate minerals. Int J Miner Proc 59(3):247–266

12. Davidovits (1994) High-alkali cements for 21st century concretes Paper presented at the V. Mohan malhotra symposium on concrete technology: past, Present and Future University of California Berkeley
13. Bureau of Indian Standards (BIS) (1992) Common burnt clay building brick specifications. IS 1077. New Delhi, India
14. Bureau of Indian Standards (BIS) (1992) Methods of tests of burnt clay building bricks-determination of compressive strength, water absorption and efflorescence [Third Revision]. IS: 3495 Part-I-III. New Delhi, India
15. Rickard WDA and van-Riessen A (2014) Performance of solid and cellular structured fly ash geopolymers exposed to a simulated fire. Cement Concr Comp 48:75–82
16. Sanjayan JG, Nazari A, Chen L, Nguyen GH (2015) Physical and mechanical properties of lightweight aerated geopolymer. Constr Build Mater 79:236–244

Metallurgical and Mining Waste Utilization in Preparation of Geo-Polymeric Bricks as the Future Construction Material

Sneha Dandekar, Kavita Deshmukh, Bhagyashree Bangalkar, Pooja Zingare, Dilip Peshwe, and Kirtikumar Randive

Abstract An excess amount of waste generated by the energy and metal industries has posed a great challenge of its disposal due to its adverse effects on the environment. There have been several initiatives world over, towards minimizing, detoxifying, and utilizing wastes, so as to make a value-addition in the existing mining operations. Among others, the geopolymers emerge as a viable, low-cost, and environment friendly solution for using industrial wastes such as fly-ash. The geo-polymers are inorganic, ceramic material, which forms a long-range, covalently-bonded, non-crystalline (amorphous) networks. The bricks manufactured from fly-ash and rice-husk silica is well established, which, on the whole, work on the same principle of amorphous silica. However, in the present study, the source of the silica has been altered by using Bamboo (Bambusa vulgaris) leaves, which is known to be a potential agro-waste and capable to producing an amorphous silica with ~92% yield. During the present study, the ternary geo-polymer brick was synthesized using fly ash, iron oxide slag (red mud) and bamboo extracted amorphous silica. The structural and mechanical properties of this new material were tested and characterized. The results have shown double fold increase in compressive strength, increase in durability, heat resistance and strength with respect to time. Therefore, the new material has a potential to revamp construction and infrastructure industries by way of providing low-cost alternative to the clay bricks as well as sand. It will also have a positive impact on legal as well as illegal sand-mining, which is one of the greatest environmental threats today.

Keywords Fly ash · Bamboo leaves · Geo-polymeric brick · Crushing strength · Heat resistance

S. Dandekar (✉) · D. Peshwe
Metallurgical and Materials Engineering Department, V.N.I.T, Nagpur, India

K. Deshmukh
Mayur Industries, Hingna, Nagpur, India

B. Bangalkar · P. Zingare
Department of Physics, RTM Nagpur University, Nagpur, India

K. Randive
Department of Geology, Rashtrasant Tukadoji Maharaj Nagpur University, Nagpur, (MH), India

K. Randive et al. (eds.), *Innovations in Sustainable Mining*, Earth and Environmental Sciences Library, https://doi.org/10.1007/978-3-030-73796-2_5

1 Introduction

Geopolymer is a material originated by inorganic poly-condensation, by so-called "geo-polymerization" in other words it is an inorganic, ceramic materials that form covalently bonded long-range amorphous network [1, 2]. Geopolymer is new and environment friendly material in concrete industry and better alternative to Ordinary Portland cement. Ordinary Portland cement is replaced by geopolymer in construction industry. Wide range of application based on various properties of geopolymer is found. It is used as heat and fire-resistant coatings and adhesives and has various application as new binders for fire-resistant fibre composites, high-temperature medicinal and encapsulation for dangerous radioactive waste and cements for concrete and so on.

Geopolymer is produced from minerals such as fly ash, clay, red mud with addition of alkaline solution. Fly ash is the waste product formed due to burning of coal in thermal power plant and red mud is generated as waste material in Aluminium industry. Manufacture of Ordinary Portland Cement (OPC) involves huge amount of energy as well as it releases enormous amount of carbon into environment and is responsible for 7% of total CO_2 released in the environment [3]. OPC is widely used in manufacturing of concrete bricks. Due to the global concern about CO_2 emissions nowadays, it is well accepted that a new material for manufacturing bricks is in need to replace concrete bricks with improved environmental, mechanical, and durability performance [4]. The alkali activated geopolymer brick is one potential alternative to the concrete bricks. It may help to address the mentioned problem once the mechanical and chemical behaviours are understood.

The chief constituents of geopolymer are silicon and aluminium which can be obtained by thermally stimulated natural materials such as kaolinite or industrial byproducts like an alkali activated solution and fly ash which polymerizes the materials and create stiff structure networks and molecular chains. Deramen [5] reviewed the mechanical properties of the geopolymeric bricks made from fly ash, bottom ash, red mud, alkaline activator and water. In present work we found that iron ore slag can also be used instead of red mud and it possesses more or less same elemental properties and characteristics.

From the environmental and ecological point of view, the disposal of this industrial and mine waste is a matter to be overlooked. Therefore, from last few decades, harmless utilization of these substances is the subject of great research. The fly ash consumption saves the disposal area and also the natural resources; apart it saves energy and protects the environment [6]. By bringing ash revolution into effect, the optimum utilization of ash and converting it into eco-friendly products is a need of the hour.

For the formation of geopolymer, silica is prepared by extraction from bamboo leaves. Bamboosa Dendrocalamus species of bamboo tree is rich in silicon content. NaOH is responsible for the leaking of Al^{3+} and Si^{4+} ions in a solution which are required for bond formation. These ions leach out and react with available oxygen to form chain like structure which looks like polymer. When NaOH concentration is low the bond formation doesn't take place as the free ions are unable to leach out, whereas at higher concentration excess alkali concentration precipitation of these ion on the surface takes place which leads to hurdle in bond formation. Hence, optimum alkali concentration becomes a very important aspect of geopolymer formulation. Also, the temperature variation affects the properties of geopolymer.

Polymerization and Microstructures

Polymerization includes a noticeably rapid chemical reaction in alkaline condition on Si-Al, resultant in formation of ring structure and three-dimensional polymeric chain. Comprising of Si-O-Al-O bonds in the form of X [-(SiO_2) z-AlO_2] n. H_2O where, X resembles alkaline element or cation such as K/Na/Ca, n is the degree of the polymerization or polycondensation and z represents the bonds number. The formation of geopolymer material can be shown as described by Eqs. (A) and (B) [7].

$$\underset{\text{(Si-Al materials)}}{n(Si_2O_5.Al_2O_2)} + \underset{\text{(Alkaline activator)}}{2nSiO_2 + 4nH_2O + NaOH/KOH} \longrightarrow Na^+, K^+ + \underset{\text{(Geo-polymer precursor)}}{n(OH)^3\text{-}Si\text{-}O\text{-}\underset{\substack{| \\ (OH)_2}}{Al^-}\text{-}O\text{-}Si\text{-}(OH)_3} \quad \textbf{(A)}$$

$$n(OH)_3\text{-}Si\text{-}O\text{-}\underset{\substack{| \\ (OH)_2}}{Al^-}\text{-}O\text{-}Si\text{-}(OH)^3 + NaOH/KOH \longrightarrow \underset{\text{(Geo-polymer backbone)}}{(Na+,K+)\text{-}(\text{-}\underset{|}{Si}\text{-}O\text{-}\underset{|}{Al^-}\text{-}O\text{-}\underset{|}{Si}\text{-}O\text{-}) + 4nH_2O} \quad \textbf{(B)}$$

Fernandez-Jimenez et al. [8] stated that, "The two steps for the chemical process of forming geopolymers are dissolution of raw materials in alkaline solution to form Si and Al gel on the materials surface and secondly, polycondensation to form networked polymeric oxide structures" [9]. The meticulous mechanism of setting and toughening of the geopolymer material is indistinct [10]. Though, most anticipated mechanism consists of the chemical reactions which may comprise of the following stages: (i) Primarily, Si and Al atoms from the parent material gets dissolved through the action of hydroxide ions, (ii) the predecessor ions get oriented or transported or get condensed into the monomers and (iii) the monomers get poly-condensed or settled or get polymerized into the polymeric structures. These stages are intersected amongst each-other and can befall almost synchronously which makes it challenging to separate and inspect them independently.

2 General Details of Materials, Processes and Parameters

The vital role in the formation of geopolymer is of the materials used. Resources enrich in Si (bamboo leaves, fly ash, rice husk, slags) and enrich in Al (red mud) are the chief necessity to endure geo-polymerization. Till date alkali silicate solutions are used to dissolve the raw materials to form the reactive predecessors for the geopolymerization process. Silicate initiation increases with the dissolution of the initial materials and initiates favourable mechanical properties [10].

2.1 *Fly Ash*

It is an environmental pollutant, generated in ample amount as a waste material of burning fossil fuel coal-based thermal power station, it has a potential to be a resource material [11]. Worldwide, annually 900 million tonnes of fly ash is produced, of which only 30–40% is being utilized for various purposes, comprising cement and concrete production [12]. Fly ash is generally of spherical shape which improves the consolidation of concrete and aligned materials [10] and size range from 2 to 10 μm. The amount and type of incombustible material determines the composition of fly ash [13]. It primarily consist of SiO_2 (59.0%), Al_2O_3 (21.0%), and Fe_2O_3 (3.7%) [14] and CaO with some amount of potassium, magnesium, sodium, sulphur and titanium. There is a wide range in the chemical composition of fly ashes representing that the coal used in the power plant all over the world is of wide variations [13].

2.2 *Sodium Hydroxide*

Amalgamation of sodium hydroxides (NaOH) or potassium hydroxide (KOH) with sodium silicate (Na_2SiO_3) or potassium silicate (K_2SiO_5) is commonly used as an alkaline activator, for the formation of geopolymers [15]. Leaching out of Al^{3+} and Si^{4+} ions are usually maximum with sodium hydroxide solution as compared to potassium hydroxide solution. Thus, significant factor to leach out silica and alumina from the fly ash particles is the alkali concentration. NaOH being colourless, odourless and has more basicity is used conveniently. Duchesne [16] states that the NaOH in the activating solution makes the gel less smooth and proceeds the reaction rapidly.

2.3 *Sodium Silicates*

Sodium silicates (Na_2SiO_3) are silicon rich white powder/crystalline solids/colourless glassy material, they are readily soluble in water generating alkaline solutions. In geopolymerization process, it is used in alkaline activation process and comprises an important process in geopolymerization technology [17].

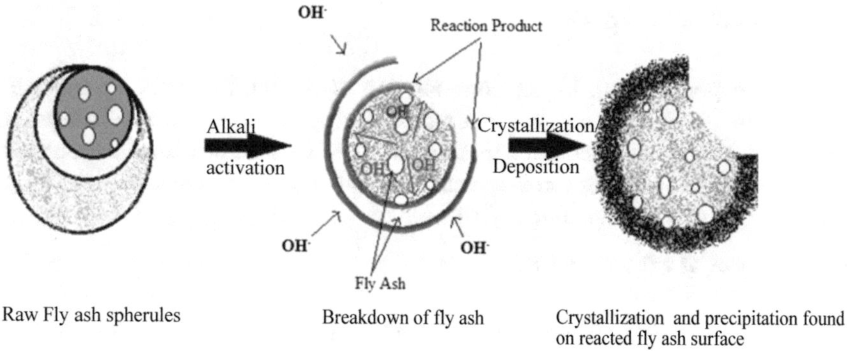

Fig. 1 Schematic showing alkali activation process [8]

2.4 Alkali Activation

Absorption of the solid material plays a significant part in the alkali activation [18]. Figure 1 illustrates the alkali activation process where Si and Al dissolution, after reaction of fly ash with alkaline solution is depicted. In the said process, the alkali attack on the surface of the particle and the higher molecules condense in a form of gel. These molecules then expands to the larger holes, exposing the hollow or partly filled smaller particles with smaller sized ashes to the alkaline attack in two-way that is from outside- in and inside-out direction. Till the ash gets nearly or completely consumed the reaction product is stimulated both inside out of the shell of the sphere [19]. The breaking down of Si–O–Si bonds and the penetration of the Al atoms results in the formation of alumina-silicate structures that is alumina-silicate gel (geopolymer precursors) due to polymerization and nucleation process represents the considerable features of this reaction. While the poly-condensation reaction, depending upon the composition of the starting material and condition of the reaction the calcium silicate hydrates and calcium aluminate hydrates phases may also be originated with the formation of secondary H_2O. Crystalline or partially amorphous or amorphous (gel-like) substances may be initiated in dependence on the character of preliminary raw materials and on the circumstances of the reaction [10].

2.5 Variation in Parameters

Aiming to study the effect of various parameters; the amount of Si, Al and concentration of alkali was varied during the study.

A. **Amount of Si and Al**

For making geopolymer, higher concentration of Al and Si atoms are required. These atoms form a polymeric chain like structure, because of which strength of the geopolymer is increased. Proper ratio of Al to Si must be maintained throughout the process of geopolymer formation and it usually lies within the scope of 2.0–3.5 to make good concrete and bricks [10].

B. **Amount of activators used**

Commonly used activator for manufacturing geopolymer is NaOH. A fixed ratio of NaOH and Sodium Silicate is added to the waste material obtained from steel industry. The total volume of activators should be properly maintained, for positive change in crushing strength of the geopolymer brick obtained [9].

By mixing strong alkalis, such as sodium hydroxide (NaOH), potassium hydroxide (KOH), sodium silicate or potassium silicate with aluminosilicate-reactive material, geopolymer is synthesized. With a strong alkali solution, aluminosilicate reactive materials dissolves and form free SiO_4 and AlO_4 tetrahedral units. For dissolving of Si and Al atoms to form geopolymer precursors, presence of Alkali activating solution is important. The behaviour of aluminium in sodium silicate ($Na_2O \cdot nSiO_2$) depends on the alkalinity of the products, i.e. on the mass ratio of SiO_2/Na_2O, usually ranges from 2 to 4. The higher the ratio, the less alkaline the product is obtained and the less action of sodium silicate is observed on aluminium [20]. Gel of $Si(OH)_4$, $Al(OH)_3$ and alumino-silicates was formed and it resulted in congealtion of solution in accordance with reduction of ions in the solution. Gel formation has two opposing effects: first, being that depletion of ions which stimulated further leach out of ions from fly ash particles and secondly, congealing resulted in low movement of solution and ions, predominantly at the surface of fly ash and leaching out of ions gets stunted. Si^{4+} and Al^{3+} species diffused into solution leading to decline in Si^{4+} and Al^{3+} concentration at fly ash particle surface and congealing of solution [21].

During the amalgamation of geopolymers, NaOH turn out to have considerable effect, on both the compressive strength and structure of geopolymers. NaOH concentration of the geopolymeric system acts on the dissolution process and on bonding of solid particles in the final structure [22]. Curing at elevated temperatures range of 40–95 °C enhanced geopolymerization and resulted into a high compressive strength of the formed geopolymer [23–26].

2.6 Curing Duration

Room temperature cured geopolymer has compressive strength analogus to that of portland cement, that is, with an increase in age the strength increases, it is the resultant of the reaction between silica and alumina in the presence of alkali ions.

Strength development at the early period of two weeks was high, and strength gain after the period of two weeks was less except for the less molar NaOH sample, which showed a high rate of strength development nearly up to a month. An increase in the molarity of NaOH clearly enhanced the strengths of the pastes. Whereas, at higher molarity the strength of paste increases, but to a smaller extent [22].

2.7 *Curing Temperature*

The mixture can be cured at room temperature or can be cured at particular temperature in furnace. High curing temperature resulted in the more compressive strength of geopolymer, but beyond 60°C, it did not led to substantial increase in the compressive strength [15].

3 Methodology

3.1 *Preparation of Amorphous Sodium Silicate from the Leaves of Bamboosa Dendrocalamus*

For the preparation of amorphous Sodium Silicate, leaves of Bamboosa Dendrocalamus were collected and dried. The 100 g of dried bamboo leaves were grinded and were placed in the furnace at 800 °C for 6 h. The sample was kept undisturbed for 24 h and then weighed out to be 17.5 g. The sample was then grinded with the help of mortar and pestle. In order to leach out silica, the sample is dissolved in 3% HCl for 24 h then weighed out again to be 15.5 g. Almost 88% yield rate of silica was observed throughout the experimentation from Bamboo ash. The sample is then filtered out and washed with distilled water and is allowed to dry. To prepare 100 ml sample, 10 M NaOH, 40 g NaOH was dissolved in 100 ml distilled water and was left undisturbed for 24 h. Then, 5 g of extracted silica was taken and dissolved in 20 ml, 10 M NaOH and was placed on magnetic stirrer for 1 h. The solution was left undisturbed for 24 h for the formation of sodium silicate.

3.2 *Formulation of Geopolymer Brick by Utilising Industrial Waste*

For the formulation of geopolymeric brick; 54 g of fly ash, 6 g of iron slag and 20 ml of sodium silicate were mixed with the help of domestic mixer. In order to maintain the consistency 10 M NaOH was added and the slurry was poured in the mould. The mould was kept undisturbed and allowed to solidify for 1 week. As the slurry was solidified into brick, it was removed from the mould and was cured at 80 °C for 24 h and the brick was exposed to ambient atmosphere for 30–40 days. Here the reaction involved is discussed above.

3.3 To Investigate the Change in Strength of Geopolymer Brick with Change in Temperature

In order to study the strength of geopolymeric brick with change in temperature the brick was cured at 60, 70, 80 and 90 °C respectively, and the strength was measured.

3.4 Testing Techniques

The XRD analysis of the prepared silica was carried out on Panalytical X'Pert Pro (model-PW 3040/60) diffractometer with Cu Kα radiation ($\lambda = 1.54$ Å) produced at voltage of 45 kV and current of 40 mA. Scanning was performed at the 2θ angle, ranging from 0 to 100° with a scan step size and time per step of 0.01° and 15 s, respectively. The morphological changes were examined using Scanning Electron microscope (SEM-JEOL 6830A). Prior to the study of surface morphology, gold coting was done over the material by means of auto sputter (JOEL-JFC 1600 auto fine coater) in order to make the material conducting for obtaining images. Fourier Transform Infrared (FTIR) spectra of the material was logged using a Perkin Elmer Spectrum. Potassium Bromide (KBr) was used for collecting the background and a total of 20 scans were recorded. The strength of the brick was measured using the Compression Testing Machine (CTM) and the compression curve at various temperature were studied.

4 Results and Discussion

The silica was formed from dried bamboo leaves. The leaves were first dried, then grinded and then kept in furnace at a temperature of 800 °C for 6 h. The silica thus obtained is later leached in HCl and is characterized using XRD, SEM and FTIR. Figure 2 shows the dried bamboo leaves used as raw material and the final silica obtained.

For the formation of brick of different NaOH molarity, the formed silica was mixed with the NaOH solution of different molarities and sodium silicate solution was synthesised. Fly ash, iron slag and the sodium silicate were mixed and the slurry was prepared. The prepared slurry was then poured in the mould of dimension 5 × 5 cm. The mould was kept undisturbed for a period of month and was cured at various temperature and brick formation takes place (Fig. 3).

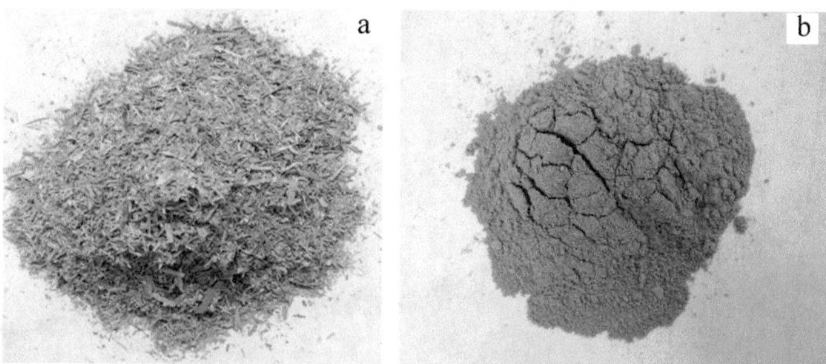

Fig. 2 **a** Dried and grinded bamboo leaves. **b** Silica obtained after the treatment of bamboo leaves

Fig. 3 Showing prepared brick

4.1 X-Ray Diffraction

In Fig. 4 the XRD pattern shows an amorphous peak at 21.8° 2θ position. An amorphous SiO_2 peak prepared by sol-gel method was also noted at $2\theta = 23°$ [27]. Similarly, Zhang et al. [28] reported a wide peak at $2\theta = 21°$, resembling the amorphous silica. This confirms that the extracted silica obtained is indeed amorphous. It is documented that for the formation of geopolymers the SiO_2 should be amorphous in nature, as the amorphous or non-crystalline SiO_2 lacks the long-range order of its structure and it helps in the bonding of NaOH to form sodium silicate solution.

4.2 Scanning Electron Microscopy

The morphology of the extracted silica was observed using Scanning Electron Microscopy. Figure 5 shows the SEM morphology of the globular shaped silica. The approximate particle size of silica observed was 5 μm (Fig. 5a). Openings can

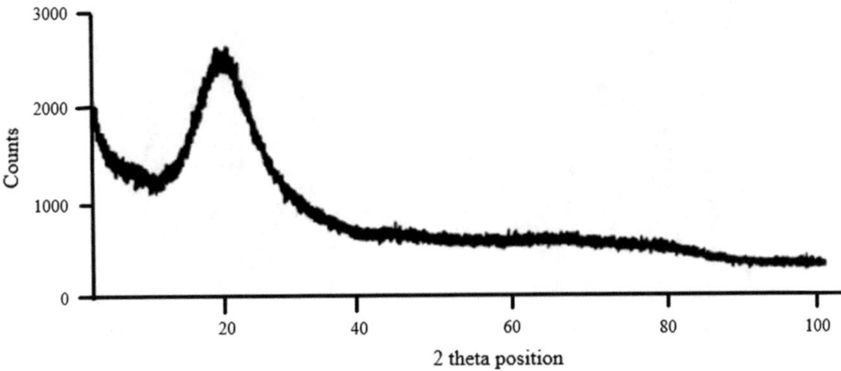

Fig. 4 XRD of treated Silica

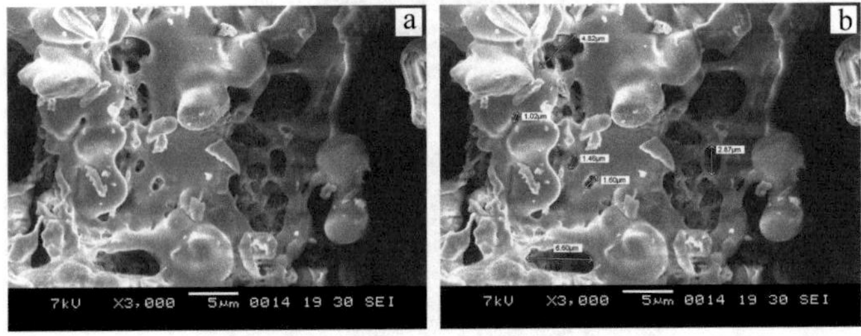

Fig. 5 (a) Showing SEM results of the Silica particle (b) Showing formation of porous silica

be observed resulting in the formation of porous silica (Fig. 5b). Apart, the pores were observed to range in size from approx. 1 to 6.06 μm. This ascertains the fact that the obtained silica is porous in nature (Fig. 5b).

4.3 Fourier Transformed Infrared Spectroscopy

Figure 6. shows the FTIR spectrum of amorphous SiO_2. The infrared (IR) band at 3636 cm^{-1} is due to presence of free –OH ions and shows stretching vibration of H_2O molecules. The band at 3246 cm^{-1} represents the –OH stretching of hydroxyl group/Si-OH group. The IR band at 1632 cm^{-1} shows bending vibration of H_2O molecules. The band at 1089 cm^{-1} denotes vibration of Si-O-Si asymmetric stretch, especially observed in long chain that determines the formation of amorphous silica. Whereas, the IR band in the range of 972–956 cm^{-1} shows stretching vibrations in silanol (Si-O-H). The band at 800 cm^{-1} shows symmetric stretching vibrations of Si-O-Si bonds and the IR band at 474 cm^{-1} is because of bending vibrations in O-Si-O

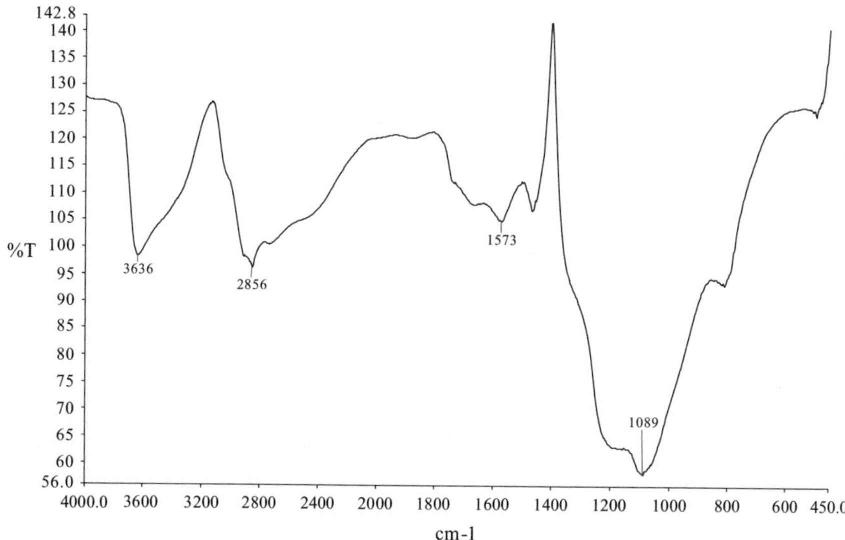

Fig. 6 Showing FTIR results of the Silica particle

bond. Hence, the FTIR investigation confirms that the amorphous SiO_2 was formed and as no other peaks were present ascertains that the extracted silica was pure.

4.4 *Compression Strength*

The compressive strength of the synthesized bricks (5 × 5 cm) were measured and the average of 3 is reported. The minimum compressive strength of brick required is 3.5 N/mm^2 when tested as per IS-3495 (Part1):1976. Moreover the minimum of 20% tolerance of compressive strength is accepted in case of any individual brick [6].

From the compression strength it was studied that, the strength of the geopolymeric brick increases with respect to curing temperature. From Fig. 7a it is noted that at 60, 70, 80 and 90 °C the strength of the brick was 2.5 MPa, 3.37 MPa, 4.1 MPa and 4.9 MPa respectively. The highest compressive strength was found to be 4.9 MPa at 90 °C with curing age of nearly a month, it is due to the ease of bond formation at higher temperature which affects the compressive strength of the brick [29]. However, from Fig. 7b, it is noted that the strength of the geopolymeric brick increases with the increase in molarity of NaOH and shows similar strength for particular molar concentration.

As at lower concentration, the leaching of Si^{+3} and Al^{+4} is less, hence forms a material of low compressive strength. It was studied that, the strength of the brick decreases at higher molar NaOH concentration (12 M), it could be due to more

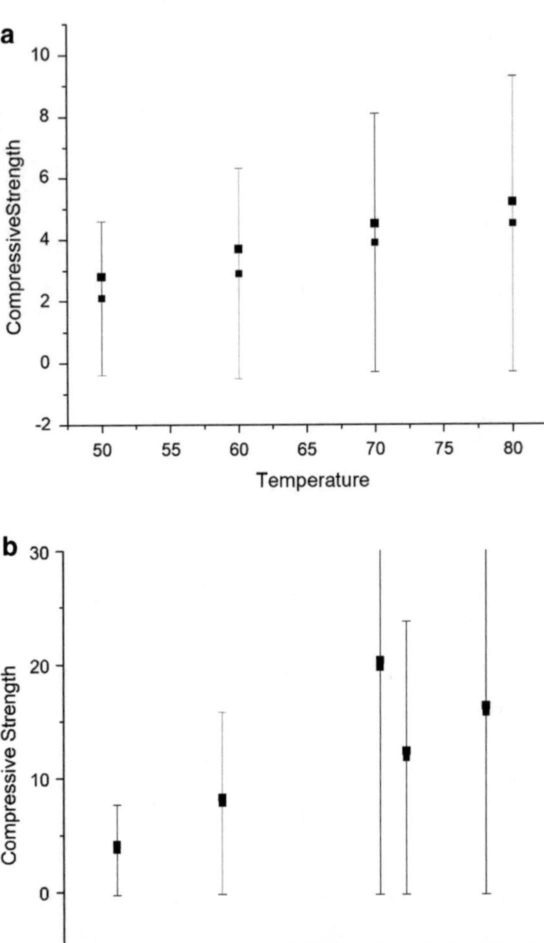

Fig. 7 **a** Showing increase in compression strength of geopolymeric bricks with respect to temperature. **b** Showing increase in compression strength of geopolymeric bricks with respect to molarity

amount of Na^+ ions existing in the cavities which avoid the formation of complete chain, therefore the bond formation is weak [29, 30].

5 Conclusion

It can be concluded from present study that geopolymeric bricks made using fly ash, red mud and silica helps in the utilization of industrial and mine wastes, thereby reducing the space to dump fly ash and red mud and avoid the leaching of these hazardous elements into the groundwater and soil. After the utilization of bamboo stem, the least used bamboo leaves can be successfully used in the production of silica. Hence, making the prepared brick completely environment friendly. The structural and mechanical properties, especially the compressive strength, of the brick is greater than the fireclay bricks available in the market with the effective price.

References

1. Davidovits J (1988) Soft mineralurgy and geopolymers. In: Proceedings 1st european conference of soft mineralurgy “Geopolymer ‘88”, Compiegne
2. Traoré S, Diarra A, Kourouma O, Traoré DL (2019) Survey of bauxite resources, alumina industry and the prospects of the production of geopolymer composites from the resulting by-product, geopolymers and other geosynthetics, survey of bauxite resources, alumina industry and the prospects of the production
3. Chen C, Habert G, Bouzidi Y, Jullien A (2010) Environmental impact of cement production: detail of the different processes and cement plant variability evaluation. J Clean Prod 18(5):478–485
4. Hasanbeigi A, Menke C, Price L (2010) The CO_2 abatement cost curve for the Thailand cement industry. J Clean Prod 18(15):1509–1518
5. Wan Mastura WI, Mustafa AM, Bakri Al, Andrei V, Kamarudin H, Ioan G, Nizar IK Aeslina IAK, Binhussain M (2014) Processing and characterization of fly ash-based geopolymer bricks. Revista De Chimi 65(11):1340–1345
6. Gawatre D, Vairagade L (2014) Int J Sci Res (IJSR) 3(10):2247–2252
7. Cao VD, Pilehvar S, Salas-Bringas C, Szczotok A.M et al (2018) Influence of microcapsule size and shell polarity on the time-dependent viscosity of geopolymer paste, industrial & engineering chemistry research. J Am Chem Soc 1–9
8. Fernandez-Jimenez A, Palomo A, Criado A (2005) Microstructure development of alkali-activated fly ash cement: a descriptive model. Cem Concr Res 35(6):1204–1209
9. Mohajerani A, Suter D, Jeffrey-Bailey T, Song T, Arulrajah A, Horpibulsuk S, Law D (2019) Recycling waste materials in geopolymer concrete. Clean Technol Envir 1–45
10. Abdullah MMA, Hussin K, Bnhussain M, Ismail KN and Ibrahim WMW (2011) Mechanism and chemical reaction of fly ash geopolymer cement-a review. Int J Pure Appl Sci Technol 6(1):35–44. ISSN 2229–6107
11. Hardjito D, Wallah SE, Sumajouw D (2004) On the development of fly ash based geopolymer concrete. Int Concr Abstr Portal Mater 101(6):467–472
12. Gamage N, Liyanage K, Fragomeni S (2011) Overview of different type of fly ash and their use as a building and construction material. In: conference: international conference of structural engineering, construction and management, Sri Lanka
13. Malhotra VM, Ramezanianpour AA (2005) Fly ash in concrete, CANMET, 1994. ISBN 0660157640, 9780660157641:307
14. Khairul NI. Kamarudin H, Mohd SI (2007) Physical, chemical & mineralogical properties of fly ash. J Nucl Sci Technol 4(2007):47–51
15. Hardjito D, Wallah SE, Sumajouw DMJ, Rangan, BV (2005) Introducing fly ash based-geopolymer concrete: manufacture and engineering properties, in our world in concrete and structures international conference, Singapore: 271-278
16. Duchesne J, Duong L, Bostrom T, Frost R (2010) Microstructure study of early in situ reaction of fly ash geopolymer observed by environmental scanning electron microscopy (ESEM). Waste Biomass Valor 1:367–377
17. Ridzuan ARM, Khairulniza AA, Arshad MF (2014) Effect of sodium silicate types on the high calcium geopolymer concrete. Mater Sci Forum 803:185–193
18. Skvara F (2007) Alkali activated material-geopolymer. Ceram Silik 51(3):173–177

19. Pacheco-Torgal F, Castro-Gomes J, Jalali S (2008) Alkali-activated binders: a review part I, historical background, terminology, reaction mechanisms and hydration products. J Constr Build Mater 22:1305–1314
20. Vargel C (2004) Inorganic salts, corrosion of aluminium. Elsevier, 17–441
21. Rattanasak U, Chindaprasirt P (2009) Influence of NaOH solution on the synthesis of fly ash geopolymer. Miner Eng Elsevier 1073–1078
22. Somna K, Jaturapitakkul C, Kajitvichyanukul P, Chindaprasirt P (2011) NaOH-activated ground fly ash geopolymer cured at ambient temperature. Fuel 90(6):2118–2124
23. Chindaprasirt P, Chareerat T, Sirivivananon V (2007) Workability and strength of coarse high calcium fly ash geopolymer. Cem Concr Compos 29:224–229
24. Swanepoel JC, Strydom CA (2002) Utilisation of fly ash in a geopolymeric material. Appl Geochem 17(8):1143–1148
25. Panias D, Giannopoulou IP, Perraki T (2007) Effect of synthesis parameters on the mechanical properties of fly ash-based geopolymers. Colloids Surf A 301:246–254
26. Bakharev T (2005) Geopolymeric materials prepared using class F fly ash and elevated temperature curing. Cem Concr Res 35:1224–1232
27. Martinez JR, Palomares S, Ortega-Zarzosa G, Ruiz F, Chumakov Y (2006) Rietveld refinement of amorphous SiO_2 prepared via sol–gel method. Mater Lett 60:3526
28. Zhang G, Xu XuY, Wang D, Xue Y, Su W (2008) Pressure-induced crystallization of amorphous SiO_2 with silicon-hydroxy group and the quick synthesis of coesite under lower temperature. High Press Res 28:641
29. Deraman LM, Abdullah MMA, Ming LY et al (2017) Mechanical properties on geopolymer brick: a review, 3rd Electronic and Green Materials International Conference 2017
30. Sukri D (2010) Investigating of compressive strength foam brickwall panel with different bonding by using stretcher and Flemish bond, University Malaysia Pahang

Beneficiation of Low-Grade Bauxite: A Case Study of Lateritic Bauxite of India

P. G. Bhukte, G. T. Daware, S. P. Masurkar, M. J. Chaddha, and Anupam Agnihotri

Abstract Lateritic bauxite are the products of intense subaerial weathering of alumina rich rocks. The resources of bauxite in India are in the order of 3850 million tones and occupy 5th position in World map. The occurrences of Laterite are widespread in various regions. The bauxite deposits/occurrences are, however, mainly located in the Eastern Ghat, Central India, West coast and Gujarat state. Bauxite deposits consist of four horizons namely Duricrust (Laterite), Bauxite, Saprolite (weathered) and parent rock. The base (parent) rock is responsible for the formation of bauxite profile and characteristics of bauxite ore varies depending on rock composition. JNARDDC has evaluated bauxite and laterite deposits of India from geology, mining, beneficiation and metallurgical point of view. The bauxite mines/deposits are associated with various low-grade materials and it remains unutilized at mine site due to their inferior composition. For utilization of these low-grade materials, quality of the same has to be improved. Keeping in view the improvement in quality of low-grade materials, beneficiation studies have been carried out at JNARDDC. The studies carried out on laterite (ferruginous, aluminous, siliceous) and low-grade bauxite indicated that the ore can be upgraded with relevance to reduction in iron oxide, silica and enrichment in alumina content. At present, India is importing raw materials required for non-metallurgical applications (refractory, abrasives and chemical, etc.). The beneficiation studies have shown great promise to be developed as a substitute for the applications requiring high-grade ores. This will have a significant effect on the life of bauxite mines as well as dwindling natural resources.

Keywords Bauxite · Laterite · Beneficiation

P. G. Bhukte (✉) · G. T. Daware · S. P. Masurkar · M. J. Chaddha · A. Agnihotri
Jawaharlal Nehru Aluminium Research Development and Design Centre (JNARDDC), Amravati Road, Wadi, Nagpur 440023, (M.S.), India
e-mail: pgbhukte@jnarddc.gov.in

M. J. Chaddha
e-mail: mjchaddha@jnarddc.gov.in

A. Agnihotri
e-mail: director@jnarddc.gov.in

K. Randive et al. (eds.), *Innovations in Sustainable Mining*, Earth and Environmental Sciences Library, https://doi.org/10.1007/978-3-030-73796-2_6

1 Introduction

The bauxite is the principal ore for the production of alumina/aluminium. The bauxite deposits are formed by intense subaerial weathering of alumina rich rock. The chemical, mineralogical and physico-mechanical properties of these bauxite widely vary depending upon the parent rock composition, mode of origin, geomorphological position, duration and age of bauxite formation. About 80% resources of bauxite in the Country are gibbsitic in nature however, occurrences of Laterite and low-grade bauxite are widespread in various parts of the country. During the process of bauxitisation, laterite/bauxite profile formed which comprises usually four horizons resting on the parent rock [1, 2]. The laterite is uppermost part, consists of hard material and dominated by ferruginous, aluminous and siliceous minerals. The thickness of laterite in profile is usually 0.5–4 m and varies from deposit to deposit. Bauxite thickness varies from 0.5 to 14 m. A transition zone (saprolite, PLK) is formed below the bauxite zone and is composed of silica and alumina bearing minerals. In most of the bauxite deposits of India, the bauxite rests on saprolite zone. In general, the thickness of saprolite zone is from 0.5 to 10 m and largely varies from deposit to deposit [3]. During the exploration/ mining of bauxite deposits, the low-grade materials (laterite, low grade bauxite, saprolite, partially lateritised khondalite) is unutilized due to their inferior composition. The optimum use of Country's resources of unutilized raw material associated with bauxite mines requires proper technical data and information in order to establish their suitability for industrial uses. In present scenario, high quality bauxite, suitable for non-metallurgical industries are fast exhausting in the country. The research work carried out at JNARDDC will facilitate in improvement of the quality of low-grade ores (laterite and low-grade bauxite) by using various beneficiation techniques proposed here.

2 Waste Generated at Bauxite Mine

During the bauxite mining, various low-grade materials such as laterite, saprolite, low grade bauxite, PLK (Partially Lateritized Khondalite/parent rocks) are generated however, these are discarded off due to their inferior composition. These materials have characteristics of low alumina, high silica and iron oxide content. Due to this inferior quality, these materials could not be used for alumina production or other industrial applications hence, remain unutilized at mine site. During the mining, protection of forest cover, top soil, OB Laterite, etc. is important for the healthy environment and sustainable development of the region [4].

3 Transforming Unutilized Materials (Mine Waste) into Resource

The mining sector is one of the important sectors that contribute to the national economy. The increasing demand for metal and minerals from various industry are expected to drive the market and the Asia Pacific region is projected to lead the mining waste management market. In World scenario China, Australia, Kazakhstan and India are the main countries contributing to the mining waste management market [5]. Keeping in view the optimum use of Country's resources of unutilized raw material associated with bauxite mines requires proper data and information in order to establish their suitability for industrial uses. In present scenario, day by day the resources which are suitable for non-metallurgical industries are exhausting and country is importing the high-grade bauxite. The National Mineral Policy (NMP-2019) has emphasized utilization of small group of deposits along with mineral wealth. The occurrences and isolated deposits of bauxite and laterite are scattered all over the Country and available for economic extraction of mineral values.

4 Scenario of Bauxite Deposits

The world production of bauxite estimated at 327 million tones. Australia continued to be the major producer and accounted for about 28% share in total production, followed by China (24%), Guinea (20%), Brazil (9%), India (7%) and Indonesia, Jamaica (3%) [6–9].

Country	Australia	China	Guinea	Brazil	India	Indonesia	Jamaica
Production (M.T)	96.00	79.00	66.18	29.00	23.68	11.00	10.10

In India consumption of non-metallurgical grade bauxite is about 8%. Cement sector is contributed about 40% followed by refractory, chemical, steel, ferro alloys, abrasive, etc. Gujarat State is the main supplier of abrasive and refractory grade bauxite. In current scenario, the proved reserves of high-grade bauxite, suitable for non-metallurgical industries (refractory, abrasive, chemical) are limited however, the large resources of metallurgical grade gibbsitic bauxite is available in east coast region. The high alumina, low iron and titania bauxite occur only in parts of Gujarat, Chhattisgarh, Maharashtra, Jharkhand and with scattered and scanty deposits in part of eastern and western region. Gujarat state is well known for high alumina bauxite suitable for non-metallurgical applications however, most of the deposits are already exhausted. Some of the high-grade bauxite deposits located in Central India are not accessible due to forest and tribal problems. On the other hand, vast good quality bauxite reserves located in eastern Ghat and coastal region, suitable for metallurgical industry, are mostly lying unused. The scarcity of high-grade bauxite in the Country can partly be resolved by making available these deposits for non-metallurgical industries and also encourage existing mines, to separate out value-added high-grade ore [10].

5 Beneficiation

The objective is to reduce iron oxide and silica content and enrichment of alumina in low grade ore for improvement in the quality.

5.1 Sampling and Characterization

For the present study, the representative samples of laterite and low-grade bauxite were collected from bauxite deposits located in West coast and east coast deposits of India. In west coast (Maharashtra) deposits, the thickness of bauxite is 2.5 m (avg). It is pisolitic in nature and contains moderate to high alumina as well as low silica [11, 12]. The weathered rock zone (saprolite) is placed below the bauxite zone which is soft in nature and cream, white, grey in colour. The laterite (overburden) occurs above the bauxite zone and it is hard, massive in nature with red, pink, and grey in colour (Fig. 1a).

The samples were crushed to −25 mm size by Jaw crusher. For the characterization studies, −74 μ size samples have been prepared by universal mill/ bond mill and thoroughly mixed using homogenizer. The representative sample was drawn by coning and quartering procedure. The chemical and mineralogical analysis of samples has been done by wet chemical method and XRD with XDB software respectively. Laterite is characterized by high iron oxide (20–45%), silica (6–15%) and low alumina (25–35%) content however, low grade bauxite is characterized by quite low alumina, low iron oxide as compared to laterite. Mineralogical analysis by XRD shows that laterite and low-grade bauxite is characterized by hematite, goethite, kaolinite and gibbsite minerals. The petrology studies indicate that laterite is pisolitic and iron minerals exhibit colloform texture. The gibbsites are cryptocrystalline and pseudomorph after plagioclase feldspars (Fig. 1b, c). The morphology studies indicated that laterite contains undeveloped crystals of gibbsite and in some laterites, it is hexagonal in shape (Fig. 1d). Our studies reveal that in most of the laterite deposits, the gibbsite is not well developed due to partial bauxitization [13]. It is observed that habit of minerals, crystal shape, etc. are not prominent in low grade ores as compared to bauxite ore [14].

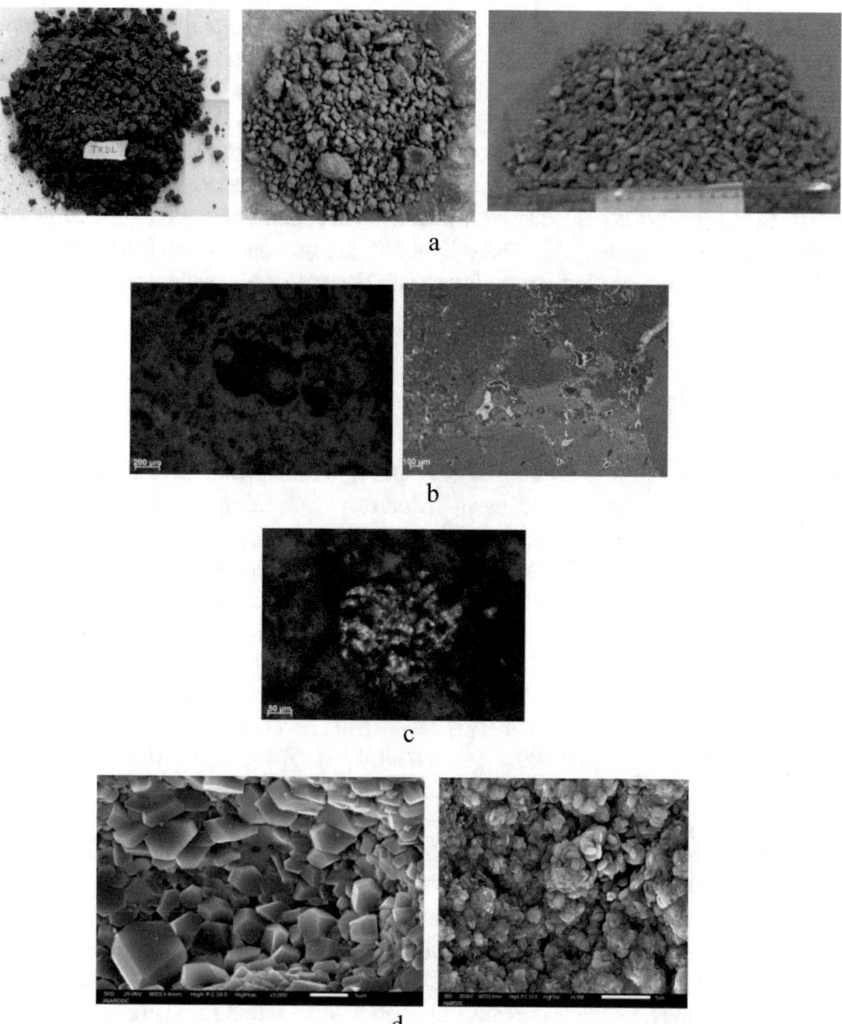

Fig. 1 **a**. Physical appearance of Laterite samples, **b**. Pisolitic texture with iron phase in Laterite, **c**. Gibbsite phase (cryptocrystalline) in low grade bauxite, **d**. Morphology of low-grade bauxite and Laterite by SEM

5.2 Beneficiation/Upgradation Studies on Laterite and Low-Grade Bauxite

The abundant resources of low-grade bauxite and laterite are characterized by high iron oxide, silica and low alumina content which restrict their use in metallurgical as well as non-metallurgical applications. Thus, any beneficiation process which can reduce silica (mainly in reactive form such as kaolinite) and iron content in bauxite and laterite, are important from processing point of view. The studies have been done by organizations/industry on beneficiation on low grade bauxite mainly for removal of impurities such as iron oxide, silica [15]. JNARDDC has done extensive studies on bauxite and laterite of various origin. Keeping in view to reduction in iron, silica and increase in alumina content, various physical beneficiation techniques such as screening, sieving, magnetic separation and hydrocyclone were adopted. The results of beneficiation studies showed that iron and silica content can be reduced from low grade bauxite and laterite [16]. The salient features and results obtained in physical beneficiation studies carried out on laterite and bauxite are given below.

5.3 Screening/Sieving

After exploration of ore, sampling of ore (sorting, braking, and screening) is important stage at mine site as well as in laboratory. Screening and sieving tests on laterite and bauxite samples from different geological origin have been carried out at JNARDDC laboratory. The clay agglomerates on the bauxite surface create lot of problems in the processing of ore. In general, its concentration in the ore is decreased by washing the ore followed by screening in which the agglomerates are detached and are removed with the action of water [17, 18]. The wet sieving has been carried out on western Ghat (Maharashtra) laterite. The sample crushed to required size (30 mm) and screened into 13 fractions namely +16, −16+11, −11+8, −8+4, −4+2, −2+1, −1+0.5, −0.5+0.25, −0.25+0.149, −0.149+0.074, −0.074+0.063, −0.063+0.045 and +0.045 mm by Anlysette vibrating screen for half an hour. The result clearly indicate that Al_2O_3 enriched above +2 mm size fraction and it increase proportional to the grain size. However, silica and iron oxide content gets enriched below −0.5 mm by wet sieving [19, 20]. The dry as well as wet sieving has been carried out on ferruginous bauxite (Panchpatmali mine). The ten different fractions clearly indicate that Al_2O_3 enriched above +2 mm size fraction and it increases proportional to the grain size. However, silica and iron content get enriched below –0.5 mm in dry screening though by wet sieving the same gets enriched below –0.063 mm (Fig. 2) [21]. The screening and sieving trials on laterite and bauxite samples of various geological origin indicated that iron content enriched in fine fractions however, increase in Al_2O_3 % particularly in coarse and middle fractions. It indicates that selective mining of a deposit can improves the overall quality of grade (ROM). The authors are in opinion that bauxite mines/deposits can be evaluated and ranked according to parameters such as bauxite resources, characteristics, present mining capacity, etc. however, quality of run of mine (ROM) and beneficiated ore are the major parameters [22]. Now a days, the said techniques are being adopted by mine owners located in Central India.

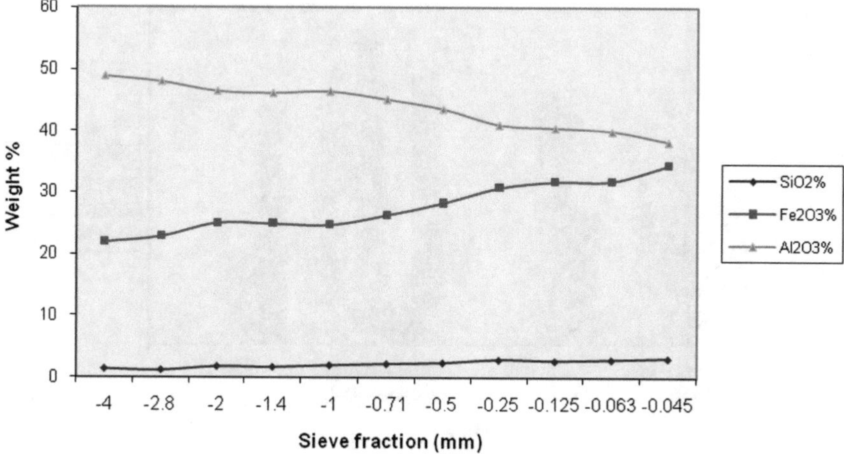

Fig. 2 Wet sieving of bauxite (East coast-Panchpatmali-Odisha)

5.4 *De-Ironing Studies by Wet High Intensity Magnetic Separator (WHIMS)*

One of the impurities in low grade ores associated with bauxite mine is high iron oxide content. The presence of high Fe_2O_3 content in ore would impact on generation of bauxite residue during alumina production. The non-metallurgical industries (refractory, abrasive, chemical) also require very less iron oxide content (Fe_2O_3 < 3%). Keeping in view the reduction of iron oxide content, tests were carried out by using wet high intensity magnetic separator (WHIMS). The trials have been done on laterite and bauxite samples with various size such as −297, −149, −74 and −44 μ and different magnetic intensity (5000 to 19,800 gauss) with 20 and 30% solids. It is observed that as magnetic intensity increases, alumina content in the non-magnetic fraction increases and iron decreases with varying recovery of ore (non-magnetic fraction).

The de-ironing tests were carried out on ferruginous laterite of western Ghats (Maharashtra) and eastern Ghats deposit by WHIMS on 10000 gauss magnetic intensity and various micron size sample. The eastern Ghats laterite is aluminous in nature and contains quite high alumina and low iron oxide compared to Western Ghats laterite. The results indicate 10–18% reduction in iron oxide and 10–15% enrichment in alumina content in non-magnetic fraction of western Ghat laterite however, eastern Ghat laterite shows considerable reduction in iron oxide (10–25%) as well as increase in alumina content by using same magnetic intensity and feed size (Fig. 3). The recovery of ore after processing in western Ghat laterite (weight % of non-magnetic fraction) is 35% however, in case of eastern Ghat sample, the recovery is 47%. It is observed that with the same magnetic intensity and laterite size, the two kinds of laterite deposit show significant difference in ore and alumina recovery as

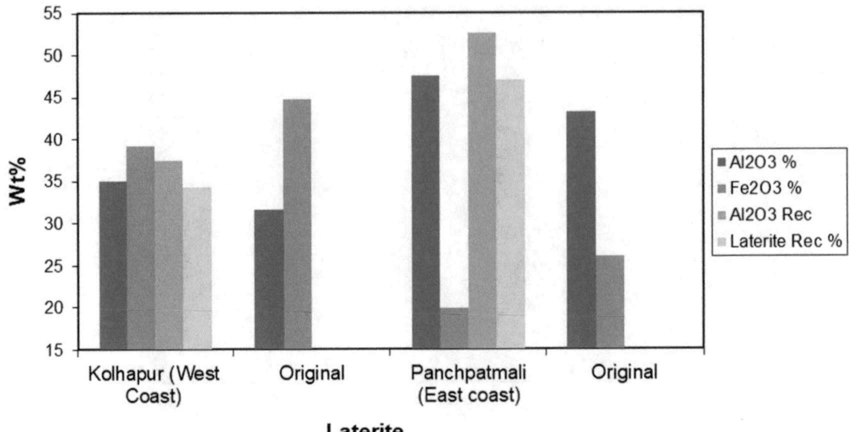

Fig. 3 De-ironing by WHIMS of laterite with 10000 gauss intensity and grain size (−297+149 μ)

well as reduction in iron content [23, 24]. The studies indicated that in beneficiation process (physical separation), magnetic intensity, grain size and ore characteristics of laterite play an important role.

5.5 *Beneficiation of Blended Bauxite by WHIMS*

An attempt has been done at JNARDDC to beneficiate various grades of overburden laterite (Panchpatmali deposit) such as ferruginous dark grey laterite, pink laterite, grey laterite and yellow laterite. The chemical composition of laterite and bauxite samples is given in (Table 1). The blended beneficiation experiments carried out at following parameters:

Sample composition :80% process bauxite and 20% ferruginous dark grey laterite
Magnetic intensity :6000, 7000 and 8000 gauss
Bauxite size :297 μ

Table 1 Chemical analysis of process bauxite, laterite and blended bauxite

Constituents (%)	Process bauxite	Ferruginous dark grey laterite	Blended bauxite
Al_2O_3	46.22	33.59	43.49
Fe_2O_3	23.10	39.14	27.28
SiO_2	3.13	4.43	3.48
TiO_2	1.90	2.04	2.11
LOI	24.70	19.25	22.47

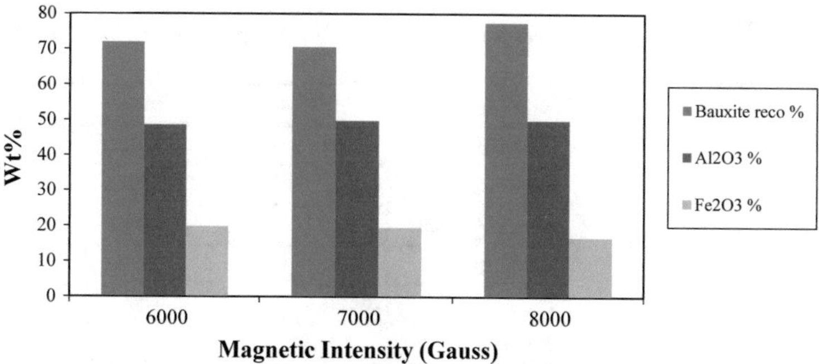

Fig. 4 De ironing of blended ore by WHIMS (−297 μ)

Bauxite quantity :50 gm
Slurry solids :30%

The tests carried out at different magnetic intensities indicate that quality of beneficiated bauxite as well as loss increases as magnetic intensity increases. However, bauxite quality in terms of alumina content is observed to be more or less same at 7000 and 8000 gauss magnetic intensity. The iron reduced in the range of 17–20% from original 27% and alumina enriched up to 50% (Fig. 4). The chemical and quantitative mineralogical composition of beneficiated bauxite (product) is presented in Tables 2 and 3 respectively. The results show that alumina loss increases with the increase in the magnetic intensities for all bauxite. The same trend is also observed with losses for iron, titania and silica bearing minerals. Not much change in losses is observed if the bauxite is ground to a finer size of 149 μ at 7000 and 8000 gauss magnetic intensity. Based on the experiments it is observed that magnetic intensity of 7000 gauss seems to be optimum for beneficiation of blended bauxite. Below 297 μ is satisfactory for beneficiation as well as for processing of the beneficiated bauxite for alumina production [25].

Table 2 Chemical composition of beneficiated product (bauxite)

Constituents (%)	Beneficiated product 1	Beneficiated product 2
Al_2O_3	49.42	48.06
Fe_2O_3	19.05	19.36
SiO_2	3.57	3.99
TiO_2	1.48	1.52
LOI	26.00	25.77

Table 3 Quantitative Mineralogical composition of beneficiated bauxite

Phases (in %)	Beneficiated product 1	Beneficiated product 2
Alumina as		
Gibbsite	44.12	43.46
Boehmite	0.85	0.42
Alumo-goethite	1.27	0.95
Kaolinite	2.76	2.76
Total	49.00	47.60
Silica as		
Kaolinite	3.26	3.26
Quartz	–	0.50
Total	3.26	3.76
Titania as		
Anatase	1.00	1.00
Rutile	–	–
Total	1.00	1.00
Iron as		
Hematite	8.00	10.00
Alumo-goethite	11.24	8.43
Siderite	–	0.34
Total	19.24	18.78

5.6 Separation Studies by Hydrocyclone Test Rig

The high amount of silica and iron oxide are the main impurities in low grade materials. For separation of impurities and upgradation of low-grade ores, study has been taken up on laterite and low-grade bauxite by using hydrocyclone test rig. The laterite sample (−297+149, −149+105 and −105+74 μ size) has been prepared for the present study. The experiments were conducted on various parameters such as solid %, pressure (psi), size of sample (micron), vortex finder (mm) and apex (mm). The results show significant enrichment of alumina (32–34%) from original 27% and reduction in iron oxide (30–24%) content from original 38% (Fig. 5). The overall recovery of laterite is in the range of 40–50%. The trials also show encouraging results with respect to reduction in SiO_2 content (9–11%) [16, 26].

5.7 Leaching Studies

The eastern Ghats and coast (Odisha & Andhra Pradesh) comprises more than 2000 million tons of metallurgical grade gibbsitic bauxite. It is characterized by

Fig. 5 Fractions of Laterite after physical beneficiation

moderate alumina, low silica and titania however, it contains high iron oxide. For non-metallurgical applications, it is necessary to bring down Fe_2O_3 content in bauxite. Keeping this in view, an attempt has been made to produce value added non-metallurgical bauxite by de-ironing of eastern ghat ferruginous bauxite (by acid leaching).

Leaching studies were carried out on ferruginous bauxite of Eastern ghat deposit with varying parameters like acid concentration, temperature, time and grain size. The bauxite is characterized by high iron oxide (Fe_2O_3−26%), alumina (Al_2O_3−44%). In general, with increase of concentration of HCl acid, the iron content reduces substantially with little loss of alumina. The various leaching tests show that iron reduction <3% can be achieved with 9–12% acid concentration at 95 °C (Table 4). It is observed that grain size of ore plays an important role during leaching studies. The maximum reduction in iron oxide content as well as enrichment of alumina achieved in −4 mm size fraction (Fig. 6). As a recovery of bauxite is concern, it is higher in −4 mm size followed by −11 mm. The test carried out with fraction of −11 mm, indicates higher recovery of ore however, insignificant reduction in iron oxide as well as increase in alumina content [21]. The experimental studies indicated that by using leaching technique non-metallurgical grade bauxite can be produced. In current scenario, India is importing high grade high alumina bauxite to fulfill the requirement for refractory industry [27].

From beneficiation studies it is observed that reduction of iron, silica content and enrichment of alumina inn low grade ore by using physical separation process. However, for substantial removal of iron and increase in alumina content, leaching process is effective. The specification required for bauxite used in metallurgical and

Table 4 Leaching by HCL

Constituent (%)	Original bauxite	De-ironed bauxite	Recovery of ore after leaching (%)
Al_2O_3	43–45	58–62	80–88
Fe_2O_3	25–26	5–2	–

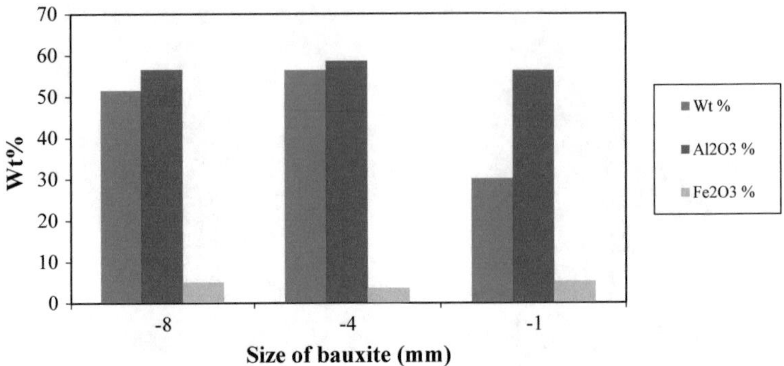

Fig. 6 Leaching (HCl) on various size of bauxite

Table 5 Specification of bauxite [28]

Grade	Major Oxides (%)			
	Al_2O_3	SiO_2	Fe_2O_3	TiO_2
Metallurgical	40–52	1.5–10	5–30	1–6
Abrasive	Min. 55	Max. 5	Max. 6	Min. 2.5
Chemical	Min. 55–58	Max. 5–12	Max. 2	0–6
Refractory	Min. 58–61	Max. 1.5–5.5	Max. 3	Max. 2.5
Low grade bauxite/Laterite after beneficiation (JNARDDC)	>58.00	<5.00	<3.00	<4.00

non-metallurgical industries is given in Table 5. By using beneficiation techniques, low grade ore can be used for non-metallurgical applications (refractive, abrasive, etc.).

6 Results and Discussion

Based on the beneficiation studies it is observed that silica and iron content enriched in fine fractions however, improvement of $Al_2O_3\%$ particularly in coarse and middle fractions. The studies on removal of impurities by WHIMS indicated reduction in iron oxide content and in this process, magnetic intensity, grain size and ore characteristics of laterite play an important role. The tests indicated that laterite can be blend with metallurgical grade bauxite for alumina production. The leaching by HCL showed that substantial reduction in iron oxide and enrichment in alumina content. By using this process, the ore required for non-metallurgical industries can be produced.

7 Summary

The quality of abundant resources of ferruginous, aluminous, siliceous laterite, low grade bauxite can be improved by using beneficiation techniques. Beneficiation studies indicated that iron oxide and silica can be significantly brought down with enrichment of alumina.

Bauxite for non-metallurgical industries meet very rigid physicochemical requirement and specifications particularly for constituents like alumina, iron oxide, titania, etc. compared to ore used for metallurgical industry. The laboratory beneficiation test results show that it is possible to produce high grade, high alumina bauxite by de-ironing, leaching process. The de-ironed laterite and low-grade bauxite may be a substitute material for high grade bauxite. After calcination, process de-ironed material can be use as calcined bauxite in refractory industry However, further detail studies are required to work out the technoeconomic of the process.

The unutilized/waste material associated with bauxite mine can be processed for value added products after beneficiation. The positive impact on cost anomaly as well as life of mine may appreciably increase by processing inferior material.

Acknowledgements The authors express their sincere thanks to S&T, Ministry of Mines, Govt. of India for sponsoring and financial support for the research projects. Authors thank Dr Anupam Agnihotri, Director JNARDDC for his constant encouragement and permission for publishing the research work. We are grateful to Dr Ashok K Nandi, President, IBAAS & Director, MIDC for giving valuable suggestions during review of paper. The valuable suggestions given by Mr G P Thakre, Manager, Castwell Industry is acknowledged. The author's thankful to Dr K R Randive, Associate Professor, Department of Geology, Nagpur University for giving suggestions about article structure and composition of the manuscript. The authors thankful to the staff of JNARDDC for carrying out laboratory work.

References

1. Bardossy G, Aleva GJJ (1990) Lateritic bauxite. Develop Econ Geol 27
2. Bhukte PG, Chaddha MJ (2014) Geotechnical evaluation of Eastern Ghats Bauxite deposits of India. J Geol Soc India 84:227–238
3. Bhukte PG, Puttewar SP, Agnihotri A (2017) Saprolite as a potential material for refractory industry. Miner Metals Rev (MMR) 113
4. Kamble PH, Bhosale SM (2019) Environmental impact of bauxite mining, a review. Int J Res Appl Sci Eng Technol (IJRASET) 7(I)
5. Research and Markets com (2018) Global mining waste management market 2017–2022 – Increasing demand for metal & minerals from the automotive industry is driving the market
6. USGS (2020) https://USGS.gov.in
7. https://aluminium.org.au/australian-bauxite/
8. https://www.miningweekly.com/
9. https://www.alcircle.com/news/brazils-bauxite
10. Bhukte PG, Puttewar SP, Daware G et al (2017) Status of lateritic bauxite deposits of India for non-metallurgical applications. J Indian Geol Congr 9(2):71–79

11. Bhukte PG, Daware GT, Masurkar SP, Mahendiran P et al (2020) Geochemical, mineralogical and petrological characteristics of Lateritic Bauxite deposits formed on Deccan Trap Basalt with reference to high-level and coastal (low level) deposits of Maharashtra. J Geol Soc India 95:587–598
12. Bhukte PG, Deshpande AS, Agnihotri A, et al (2018) Significance of geo-informatics technology in the evaluation of lateritic bauxite deposits. IBAAS Binder 30–40
13. Bhukte PG, Puttewar SP, Agnihotri A (2019) Prospects of bauxite & laterite deposits of karnataka. In: Conference "Mining in Karnataka" -Raw material security for sustainable operations organized by The Associated Chambers of Commerce and Industry of India (ASSOCHAM), pp 43–52
14. Bhukte PG, Daware G, Puttewar SP et al (2017b) A process for development of refractory aggregates from Saprolite. In: 21st international conference on non-ferrous minerals and metals (ICNFMM-2017), pp 54–59
15. Wang P, Wei D (2012) Study of the beneficiation technology for low-grade Bauxite. Adv Mater Res 454:299–304
16. JNARDDC (2016) Report on Up-gradation and utilization of east and west coast laterite. Ministry of Mines, GOI
17. Ahmad I, Hussain S, Qadir A, Khan NM (2018) Estimation of cleaning efficiency of Clay removal from Bauxite. Int J Econ Environ Geol 9(2):35–39
18. Bhukte P, Najar M, Daware, GT et al (2019) Technical assessment of Lateritic Bauxite−A case study of high-level Bauxite deposits in Maharashtra. In: 23rd international conference on non-ferrous minerals and metals (ICNFMM-2019), pp 35–42
19. JNARDDC (1999) Report on Bauxite technical databank (Madhya Pradesh, Chhattisgarh & Maharashtra). Ministry of Mines, GOI
20. Nandi AK, Bhukte P, Muthuraman K, Iyer RV et al (1999) Re-examination of bauxite occurrences of Maharashtra-Proposal for industrial use. In: Proceedings of national seminar on bauxite vision–2050. mining engineers association of India (Belgaum Chapter) Kolhapur (Maharashtra), pp 44–54
21. Bhukte PG, Puttewar SP (2014) Significance of grain size of Bauxite and Laterite during Beneficiation, Technological improvements & market developments in Aluminium Industry with special reference to value added products of Bauxite, Alumina and Aluminium (IBAAS-2014)
22. Nandi A (2019) Ranking of world bauxite mines. In: Proceedings 8th international IBAAS-GAMI binder, vol VIII, pp 14–34
23. JNARDDC (2005) Report on De-ironing of eastern ghat bauxite. Ministry of Mines, GOI
24. Bhukte PG, Puttewar SP, Agnihotri A (2017) Evaluation and beneficiation of Lateritic Bauxite deposits of India. J Geosci Res 1:251–256
25. Bhukte PG, Puttewar SP, Gundewar CS (2013) Up-gradation of bauxite for non-metallurgical applications. Int Sem Miner Proc Technol MPT 569–573
26. Bhukte PG, Puttewar SP, Daware G et al (2016) Characteristics of west coast (Goa) Laterite and possible techniques for up gradation. Section I 5:19–26
27. Bhukte PG, Puttewar SP, Chaddha MJ et al (2013) Laterite/Low grade Bauxite resources-industrial prospects. In: National conference on challenges in 21st century mining- environment & allied issues (MEAI 2013), pp 78–85
28. Bhukte P, Chaddha M, Najar M et al (2020) Beneficiation Aspects of Low-grade Unutilized Materials (Partially Lateritised Khondalite and Laterite) Associated with Bauxite Mine TRAVAUX 49. In: Proceedings of the 38th international ICSOBA conference, vol 45, no. 49, pp 87–96

Bauxite Beneficiation: An Approach to Value Addition in Mining

Basudeb Datta and Ashok Nandi

Abstract The main objective of bauxite beneficiation is to lower the concentration of reactive silica, in the form of kaolinitic clay, and also to increase the alumina content. Additionally, reduction in iron content, particularly in the form of goethite, and reduction of organic carbon can also increase the value of the bauxite product. Metallurgical bauxite is a low-value bulk ore, and therefore preference is often given to simple beneficiation techniques, such as crushing followed by dry screening or washing and scrubbing processes. The most effective way to remove both kaolinite and goethite is by washing, followed by attrition scrubbing and desliming through removal of clay-rich super fines in hydro-cyclone. Processing of bauxite for smelter grade alumina is quite sensitive to bauxite recovery and total cost for processing of ore, particularly in wet beneficiation. On the other hand, owing to homogeneous nature and consistent quality, some alumina refineries prefer washed ore compared to simple crushed run-of-mine (ROM) bauxite. Non-metallurgical bauxite ore is generally sold at higher prices, and consequently can offset lower recovery and higher beneficiation costs. The primary objective of beneficiation for non-metallurgical bauxite is to lower iron and titanium contents in the product, and also to enhance alumina values. This often requires more complicated processes, such as the use of a high-intensity wet magnetic separator, with fine grinding of bauxite to liberate the iron minerals. In some cases, reduction roasting followed by magnetic separation may be effective, depending on chemical and mineralogical characteristics of the bauxite. An acid leaching process can also be employed on physically beneficiated bauxite to lower the iron content to the level of 2–3% Fe_2O_3.

Keywords Beneficiation · Bauxite · Kaolinite · Reactive silica · Boehmite · Goethite · Washing · Scrubbing · Reduction roasting

B. Datta
Sierra Mineral Holdings 1 Ltd. (Vimetco), Freetown, Sierra Leone

A. Nandi (✉)
Mineral Information & Development Centre (I) Pvt. Ltd., Nagpur, India

K. Randive et al. (eds.), *Innovations in Sustainable Mining*, Earth and Environmental Sciences Library, https://doi.org/10.1007/978-3-030-73796-2_7

1 Introduction

As the resources of high-grade bauxite in the world are depleting, it has become almost pre-requisite for some of the mines to upgrade the ore, particularly for export, where the long and costly transportation are involved. However, the pit head alumina refineries using local bauxite may not require complicated upgrading as the Bayer process itself, is a chemical beneficiation technology. On the other hand, the beneficiation process, particularly the washing techniques, produce one of the best and consistent grade metallurgical grade bauxite, preferred by the alumina refineries compared to Direct Shipping Ore (DSO) of similar chemical characteristic. Alumina plants, in general, prefer the washed bauxite, for example MRN Trombetas (Brazil) and SMHL (Sierra Leone), despite comparatively higher cost. The washing processes are mainly adopted for high silica ores, whereas naturally occurring low silica bauxite of Guinea are used as DSO. While there is no washing in these mines, some processing such as blending on stockpiles, crushing, drying, and homogenization via stacker/reclaimer are required before exportation.

Any beneficiation process, which can effectively reduce the silica content of bauxite (mainly in reactive form such as kaolinite) is given the first priority in the metallurgical industry. Second priority is given to beneficiation methods for the reduction of iron content (e.g. goethite), which mainly causes settling issues and higher ore consumption in alumina plants. Any reduction in boehmite and organic contents of bauxite are always welcome in a low temperature alumina refinery. An important issue in the beneficiation of bauxite is the fine dispersion of iron oxides, alumina, kaolinitic clays and other minerals. However, in many—but certainly not all—deposits, the detrimental kaolinite is present in a large proportion as loose accumulations of clay-sized ($<4\ \mu m$) particles, not intergrown with gibbsite. Gibbsite is commonly present in concretions or crystals of variable size, but seldom as small as the clay-sized kaolinite. Therefore, this reactive form of silica in bauxite can be partly eliminated by adopting suitable dry and/or wet scrubbing and screening methods. In some cases, high concentrations of finely dispersed quartz can also be eliminated from run of mine (ROM) ore by dry and wet processes, to increase the value of the product.

In the case of the non-metallurgical bauxite industry, the main objectives are to lower the iron and titania content, and significantly increase alumina concentration. Given higher product prices compared to metallurgical grade bauxite, the economics are more supportive of absorbing the higher full cost of beneficiation. Due to the acute shortage of calcined bauxite for the refractory and abrasive industries, it has become necessary to develop processes for lowering the iron content and increasing alumina in metallurgical grade bauxite ore, which is more widespread in occurrence. Naturally occurring low-iron and low-titania bauxite is mainly available in Guyana, China and Suriname (for coastal plain or lowland bauxites), however, these high-grade resources are fast depleting, and China is now struggling to feed its own industries. An intensive R&D effort is required in this direction and some suggestions are put forward in this paper.

Table 1 Size and chemical assay of lateritic bauxite

Size mm	Al_2O_3 %	SiO_2 %	R–SiO_2 %	Fe_2O_3 %	TiO_2 %	LOI
+40	45.7	4.76	3.95	20.0	5.24	23.9
−40 + 30	48.5	2.74	2.49	17.6	5.06	25.8
−30 + 20	50.1	3.35	2.18	15.9	5.57	25.3
−20 + 10	46.0	4.43	3.72	20.7	3.60	24.7
+10 + 5.6	45.2	4.73	3.91	21.4	4.12	24.0
−5.6 + 3.35	43.5	5.69	4.68	23.2	4.26	22.9
−3.35 + 2	42.8	6.05	4.82	23.9	4.36	22.3
−2 + 1	43.6	6.86	5.41	24.3	4.54	20.2
−1 + 0.5	41.9	8.73	6.24	22.3	4.28	22.1
−0.5 + 0.25	39.5	12.69	8.90	22.2	4.30	20.7
−0.25 + 0.15	39.5	13.01	9.10	22.0	4.60	20.5
−0.15 + 0.1	39.4	11.90	7.80	23.4	4.59	20.3
−0.1 + 0.075	39.7	11.25	7.50	23.5	4.50	20.4
−0.075 + 0.045	39.4	11.50	7.60	23.7	4.38	20.6
−0.045	35.9	17.65	13.10	20.3	4.29	17.6

2 Bauxite Beneficiation for the Metallurgical Industry

In case of lateritic bauxite, it is common that alumina is concentrated in coarser sized particles compared to the finer sizes. Whereas the reactive silica in the form of kaolinite (clay) gets concentrated in the fine (<2 mm) and superfine (<0.05 mm) fractions. This is well-illustrated in the case of one of the lateritic bauxites of India, as shown below [1] (Table 1).

Keeping the above characteristics in view, several west coast lateritic bauxite mines of India employ dry crushing−screening process to bring down silica content from the metallurgical grade bauxite [2]. Recently, a bauxite mine of Guinea known as AGB2A has also adopted the dry beneficiation procedure to cut down silica from about 7% to below 3%, to make this one the best metallurgical grade bauxite of Guinea.

3 Wet Beneficiation of Bauxite

Microscopic examination of typical lateritic bauxite shows that mostly kaolinite and goethite form the groundmass, in-filling the larger grains of gibbsite and hematite, although part of the alumina can also be present as amorphous masses, or poorly crystalline forms in the fine-grained Al–Fe–Si matrix in bauxite. In certain bauxites, removal of clay is only possible by attrition scrubbing and wet screening after crushing to a particular size [3]. Keeping in view these general characteristics of lateritic bauxite, most of the high-silica raw ore of Brazil [4], Indonesia and Sierra Leone are beneficiated by washing, attrition scrubbing followed by screening, and finally desliming in hydro-cyclones to recover a part of good bauxite and eliminate super-fines as cyclone overflow [5]. A general flowsheet of bauxite washing is shown here in Fig. 1.

Depending on the particle size distributions and washing characteristics of individual bauxites, some of the coarse higher grade dry or wet fractions can be removed before the attrition scrubber. In order to save water and avoid washing all of the bauxite, it can be feasible to extract coarse dry fractions as such before introducing any water. Depending upon particle size distribution of run of mines bauxite, the above process flow can be further customised and optimised by feeding only the oversize (>80 mm) fraction into the crusher, and undersize directly into the attrition scrubber. As illustrated in Table 2, the dry screening process produces fairly low

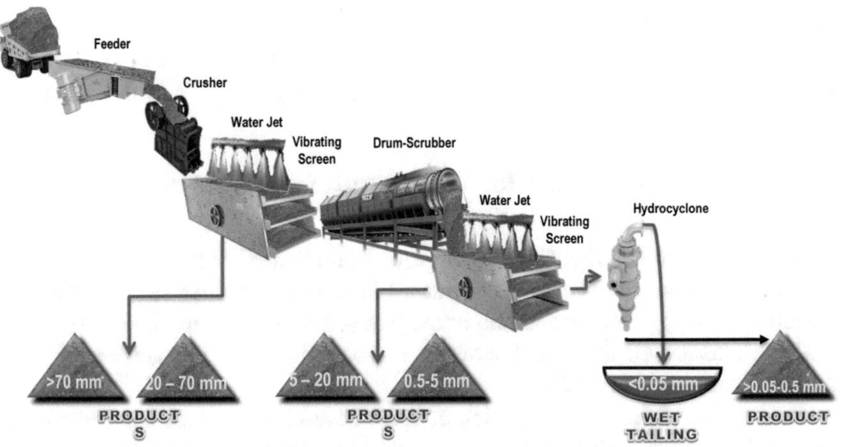

Fig. 1 General industrial flowsheet for removal of silica from Bauxite

Table 2 Characteristics of Sierra Leone bauxite (dry sieving)

Sample Fraction	$\%Al_2O_3$	$\%SiO_2$	$\%Fe_2O_3$	%LOI	Fraction %
+100 mm	47.30	3.80	20.40	25.60	20.16
−100/+ 20 mm	45.50	3.70	22.40	24.98	23.15
−20/+ 10 mm	40.90	6.10	27.40	23.62	9.68
−10/+ 6.3 mm	37.40	6.60	33.60	22.37	14.23
−6.3/+ 3 mm	36.90	5.90	35.50	21.91	7.36
−3 mm	28.70	11.60	37.20	19.96	25.42
Weighted average	39.36	6.54	28.80	23.10	100.00

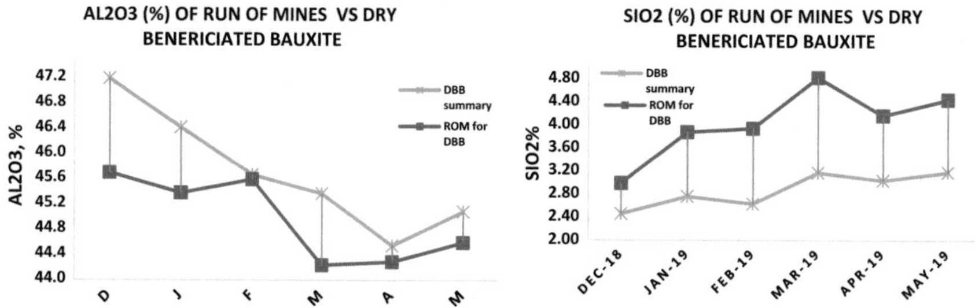

Fig. 2 Increase in alumina and reduction in silica during dry crushing–screening process

silica ore of above 20 mm fractions, and it may not be necessary to wash +20 mm size fractions [2].

Washing of bauxite generally brings down the kaolinite content, which predominates (about 80% of total silica) in the soft bauxite of Sierra Leone, and consequently alumina increases. Iron, mainly in the form of goethite & amorphous masses also decreases in the washed product, which significantly improves the settling characteristics of ore in the downstream processes. In most of the lateritic bauxites of the world, alumina and iron contents have inverse relationships and any process which decreases Fe_2O_3 proportionately increases Al_2O_3. In the case of bauxite washing, where silica is significantly decreased in the washed product, a part of the benefit arises from increasing the alumina content, however, iron may also increase or decrease depending on the nature of goethite and hematite minerals, and their ratio in the ore. Goethite, which mostly occurs as fine-grained amorphous masses, behaves like kaolinite during the washing process, and concentrates in the fines and super-fines fractions.

As indicated, dry crushing/screening processes can improve bauxite quality, however, these benefits are not significant compared to the grade improvements achieved by wet processes, as shown in Figs. 2 and 3, of Gondama bauxite of Sierra Leone.

Table 3 shows the mineralogy of typical un-washed and washed samples of Gondama bauxite of Sierra Leone; however, they are not related.

As observed in the above table, raw bauxite of Gondama, Sierra Leone has high kaolinite, quartz and goethite contents compared to the washed product. It is also reported that most of the fine, amorphous phases (Al–Si–Fe matrix) of bauxite are washed out during the intense wet beneficiation process, and overall improves the desilication, digestion and settling characteristics of bauxite during the Bayer process.

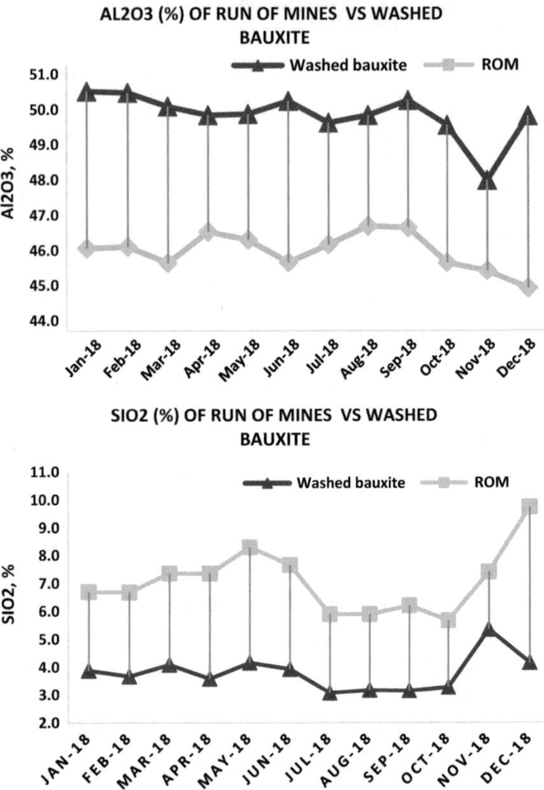

Fig. 3 Increase in alumina and reduction in silica during wet beneficiation process

4 How Attrition Scrubbing and Desliming Assists in Wet Beneficiation Process?

In order to measure the chemical and mineralogical concentrations in tailings, tests were conducted in the well-equipped laboratory of AKW in Germany, by fractional analysis, and the same samples have undergone attrition scrubbing [6]. The salient results are depicted in Tables 4 and 5.

Table 4 shows significant concentration of kaolinite in super-fines of <25 μ, where as goethite and quartz have also concentrated in the finer fractions. Gibbsite, as expected, is fairly high in the coarser particles, and slowly depletes in the finer fractions of the same bauxite. Goethite tends to concentrate in the fines; however, hematite behaves similar to gibbsite. LOI values demonstrate decreasing available alumina from coarse to fine fractions. Here, the old tailings of Gondama project of SMHL, Sierra Leone can be effectively beneficiated by eliminating below 500 μ high silica and low alumina materials. Further, these tailings were submitted to attrition scrubbing and desliming in the same AKW laboratory [3], and results are compared with simple coarse and below 500 μ fractions, as shown in Table 5.

Table 3 Mineralogy of typical Un-washed and Washed bauxite of gondama, Sierra Leone

Un-washed

Phase %	SUM	Gibs	Boeh	KaoT	Quar	Hema	Goet	Ilme	Anat	Ruti	Chem	Diff
	99.40	56.00	0.00	10.50	2.00	2.50	23.50	4.50	0.30	0.10	Anal	
Fe_2O_3 %	23.88	–	–	–	–	2.50	19.02	2.37	–	–	23.88	0.00
TiO_2%	2.77	–	–	–	–	–	–	2.37	0.30	0.10	2.93	0.16
SiO_2 %	6.89	–	–	4.89	2.00	–	–	–	–	–	6.81	−0.08
Al_2O_3 %	42.73	36.60	–	4.15	–	–	1.99	–	–	–	42.02	−0.71
H_2O%	23.36	19.40	–	1.47	–	–	2.50	–	–	–	–	–
LOI%	23.36	19.40	0.00	1.47	0.00	0.00	2.50	0.00	0.00	0.00	23.74	0.38

Washed

Phase %	SUM	Gibs	KaoT	Goet	Hema	Anat	Ruti	Ilme	Quar	Chem	Diff
	99.30	71.50	6.00	17.00	2.50	0.20	0.20	1.50	0.40	Anal	
Fe_2O_3 %	16.45	–	–	13.16	2.50	–	–	0.79	–	16.21	−0.24
TiO_2%	1.19	–	–	–	–	0.20	0.20	0.79	–	1.22	0.03
SiO_2 %	3.19	–	2.79	–	–	–	–	–	0.40	3.20	0.01
Al_2O_3 %	51.11	46.73	2.37	2.01	–	–	–	–	–	50.55	−0.56
H_2O%	27.45	24.77	0.84	1.84	–	–	–	–	–	–	–
LOI%	27.44	24.77	0.84	1.84	0.00	0.00	0.00	0.00	0.00	28.40	0.96

Table 4 Fraction analysis of bauxite tailings of Sierra Leone

Weight-%	Raw tailings	Fraction >1000 μm	Fraction 500–1000 μm	Fraction 250–500 μm	Fraction 90–250 μm	Fraction 25–90 μm	Fraction <25 μm
Mass% of fraction	100.0	40.6	13.3	9.2	12.5	9.0	15.4
SiO2	12.1	4.1	9.9	16.8	20.8	21.2	25.0
Al2O3	41.2	47.6	45.4	38.6	35.2	35.2	34.3
Fe2O3	21.1	19.6	17.4	20.7	21.3	22.1	20.5
TiO2	2.01	1.1	1.7	2.7	3.1	2.5	1.6
LOI	23.1	27.1	25.1	20.6	18.9	18.4	18.0
Gibbsite	50	68	59.5	42.5	37.5	30.5	27
Goethite	20	19	17	20	20	23	20.5
Hematite	2.5	2.5	2	2.5	3	1	2
Kaolinite	20	7.5	16	27	27	38	42
Quartz	2.5	0.5	2	4	8	3.5	5

Table 5 Comparative assays in fractions and attritioned and deslimed bauxite samples

Weight-%	Fraction >1000 μm	Fraction >1000 μm attritioned–deslimed	Fraction 500–1000 μm	Fraction 500–1000μm attritioned–deslimed	Fraction 25–500 μm	Fraction 25–500 μm attritioned–deslimed
Mass% of fraction	40.6	32.1	13.3	12.2	30.7	26.5
SiO2	4.1	3.4	9.9	7.1	19.7	16.9
A12O3	47.6	49.1	45.4	47	36.2	35.4
Fe2O3	19.6	18	17.4	17.3	21.3	23.3
TiO2	1.1	1	1.7	1.9	2.8	4.4
LOI	27.1	28.1	25.1	26.2	19.2	19.4
Gibbsite	68	73.5	59.5	69	36.9	41
Goethite	19	15	17	16.5	20.9	22
Hematite	2.5	4.5	2	2	2.3	3
Kao1inite	7.5	2.5	16	5	30.2	21
Quartz	0.5	2	2	4.5	5.5	7

Here, after the attrition and de-sliming, kaolinite along with goethite have further washed out from the coarser fractions, and deleterious contents have concentrated in the fines thanks to scrubbing in the laboratory. As a result, the quality of the +500 μ product has significantly improved in terms of silica and alumina. Based on this work, two small plants to process and recover bauxite from old tailings (waste materials of mining and earlier beneficiation), as well as fresh tailings from the bauxite washing plant have been installed at Gondama SMHL bauxite operation in Sierra Leone to recover bauxite (Fig. 4). A picture of the bauxite beneficiation plant to process the tailing materials of SMHL, Sierra Leone is shown in Fig. 5. A typical photograph of the hydrocyclone [7] for removal of clay rich fines as overflow from bauxite slurry is given here, as Fig. 6.

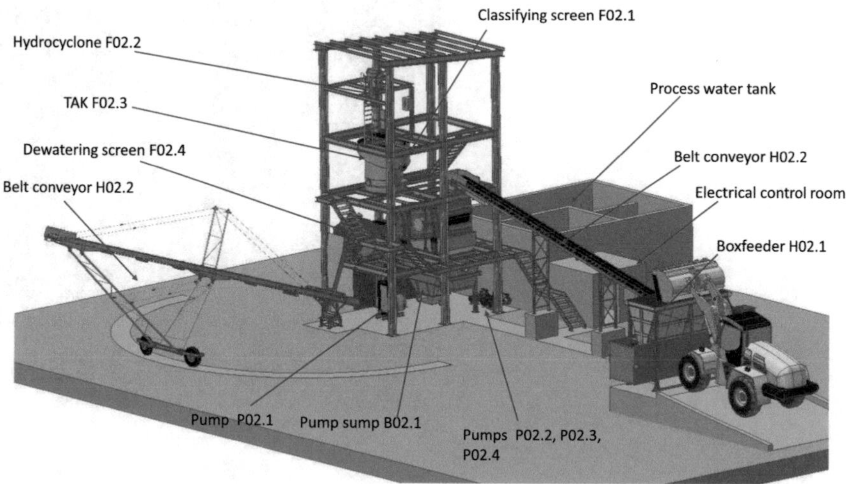

Fig. 4 Typical plant design to recover bauxite from tailings

Fig. 5 Tailing beneficiation plant of SMHL, Sierra Leone

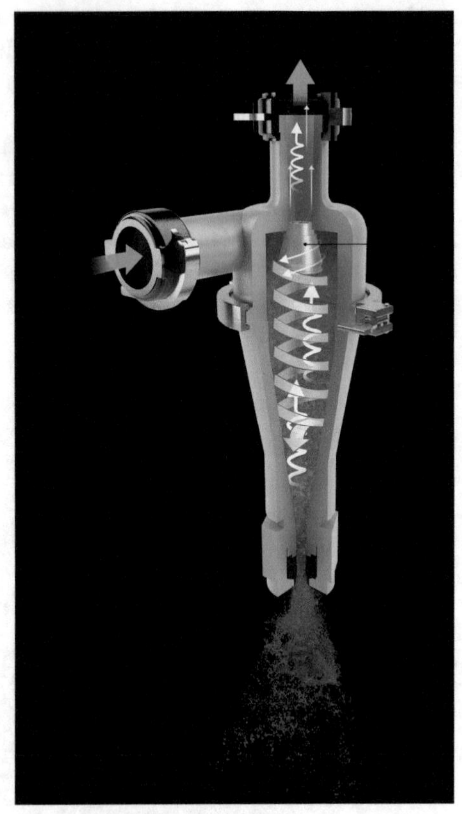

Fig. 6 A typical hydro-cyclone for separation of clay rich fines as overflow [7]

5 Bauxite Recovery

Bauxite recoveries in modern washing plants is in the range of 60–80%, however, in some of the plants, the recovery is found to be as low as 50%, as they use simple vibrating screens in place of hydro-cyclones. The quality of input bauxite, particularly silica content and targeted washed bauxite quality, dictate the washed ore recovery. The following methods may improve the bauxite recovery in the washing plants:

- Extract the good quality bauxite lumps by dry screening, before feeding material into the washing plant. As demonstrated above in Table 2, in the case of Sierra Leone bauxite, it is feasible to feed only the <20 mm fraction and improve the overall recovery.
- Wash good quality lumps to remove fine kaolinite and goethite sticking on the surface of lumps, and feed only below 30 or 20 mm into the drum scrubber, to reduce the load and increase the overall recovery.
- In the older plants, install hydro-cyclones to recover fine bauxite from live tailings, as demonstrated above.

Each bauxite has its own peculiarities and characteristics, and as such it is necessary to establish the mode of beneficiation by conducting a series of experiments at the site and laboratory pilot plant tests. Some bauxite responds quite well in the dry crushing-screening process, and for others it is necessary to use water and wash to reduce deleterious contents. Our experience shows that in low silica gibbsitic bauxite of Guinea, it is sometimes beneficial to remove fines of below 5–10 mm sizes by dry processes, to decrease the goethite & boehmite contents and improve moisture before export. This process may be useful for those deposits/mines having low alumina and high goethite content.

6 Why Alumina Refineries Prefer to Import Washed Bauxite

One of the biggest cost components of alumina refining can be the raw material cost, so having low-cost bauxite on a delivered basis is the most important. However, alumina refineries dependent on imported bauxite prefer the washed beneficiated ore compared to similar quality DSO for processing, despite comparatively higher prices [2]. Bauxite wet beneficiation requires large capital investments in a washing plant, tailings dams, power generation, water treatment, etc. and increased operating costs, environmental licensing, restrictions on water usage and rehabilitation of tailings. However, the improved bauxite quality fetches a price premium driven by refinery operating cost improvements, that can be enough to offset all these additional costs. There are some distinct advantages of processing washed bauxite, particularly for the low temperature alumina refineries using gibbsitic bauxite for sweetener in the high temperature plants. Some of the positive features of washed bauxite are as follows:

- The washed and scrubbed bauxite tends to be more uniform/homogenous in quality compared to DSO of similar grade, due to additional mixing and blending of ore.
- The washed ores have comparatively lower goethite content and also the hematite:goethite ratio improves compared to normal lateritic bauxite, which helps in better settling in the alumina plant.
- The reduction of fine-grained amorphous contents, organics and other deleterious elements of washed ore results in better bauxite digestion, precipitation and settling performance in the refinery, as well as improved alumina product quality.
- Due to the lower content of fines, handling characteristics of washed bauxite improves significantly.
- Some of the washed bauxites, such as MRN Trombetas ore, is quite reactive and gibbsite is quickly dissolved in the refinery compared to other gibbsitic DSO.
- Washed bauxite of Sierra Leone and Brazil are quite soft, and the energy requirement in grinding (Bond Work Index) of this ore is relatively low, at less than 9 KWh per tonne, compared to typical Guinea bauxite.

Keeping the above points in view, washed bauxite is considered premium grade ore in the world market for the alumina refineries However, the feasibility of washing process for each type of bauxite must be determined on a case-by-case basis and CAPEX–OPEX must justify the investment. The other advantages of bauxite beneficiation are listed below:

- Bauxite washing can significantly reduce the impurities in relatively lower grade bauxites, which would otherwise not be exportable and/or accepted in the downstream processes, to an industry-accepted quality standard, and consequently increase the conversion rate of mineral resources to ore reserves.
- Beneficiated bauxite, through improvement of specific consumption of bauxite in the downstream processes, reduces the generation of bauxite residue/red mud, whose safe storage and disposal is of major environmental concern in the industry. This helps to further cut down on specific consumption of other raw materials including caustic, lime etc., and the reduction of energy consumption, which indirectly reduces the environmental footprint during the processes.
- Majorities of the impurities, including clay and gangue minerals, which can be and should be eliminated through simple cost-effective and environmentally friendly beneficiation processes at mine sites, instead of removing them in the alumina plants at higher cost of caustic consumption, energy, storage and disposal of red mud, etc.
- The demand for imported bauxite has increased, primarily driven by China and to a lesser extent India, and beneficiated bauxite importation reduces the freight cost per ton of production of metal, which directly or indirectly reduces the environmental footprint through limiting emissions of SOX and NOX during long sea transportation of bauxite cargoes.
- Selection and introduction of the appropriate beneficiation process technology is the key in the optimization of the whole value chain of bauxite, alumina and aluminium production. Beneficiation process flowsheets can be customized and amended, based on various run of mine bauxite feed characteristics.

7 Bauxite Beneficiation for Non-metallurgical Industry

In terms of grades suitable for refractory/abrasive manufacturers, there are relatively few deposits and producers in the world due to the stringent specifications attached to refractory grade material: alumina should be minimum 55%, iron oxide levels must be lower than 2.5%, compared with ten times that for metallurgical grades [8]. However, proppants (sintered bauxite) for hydrofracturing in oil and gas wells can accept bauxite with 7–8% Fe_2O_3. Deposits which can satisfy these requirements are not widespread; hence there is relative scarcity of refractory grade bauxite sources. The general specification of raw bauxite for various non-metallurgical applications are given in Table 6 [9].

In order to produce low iron bauxite for the value-added refractory and abrasive industries, it has become necessary to beneficiate existing metallurgical grade ore,

Table 6 Chemical specifications of typical raw bauxite for various uses

Grade	Major Oxides%			
	Al_2O_3	SiO_2	Fe_2O_3	TiO_2
Metallurgical	40–52	1.5–10	6–30	1–9
Cement as additive	45–55	Max.6	20–30	3
Abrasive	Min.55	Max.5	Max.6	Min.2.5
Chemical	55–58	5–12	Max.2	1–6
Refractory	Min.55	Max.5.5	Max.3	Max.2.5
Proppants	Min.55	Max.6	Max.7	Max.8

which is available in abundance in the world [10]. Some of the processes suggested for the bauxite beneficiation for refractory/abrasive industries are highlighted below:

7.1 Reduction Roasting Followed by Magnetic Separation

In order to lower the iron content in bauxite, processes have been developed for reduction roasting, followed by magnetic separation technology, to process metallurgical bauxite. It is shown that iron and aluminium can be separated after reduction at a particular temperature (say 500–600 °C), with the help of special additives, such as coke. A metallic iron concentrate with high-iron grade, and a non-magnetic product with high alumina content, can be obtained from high-iron gibbsitic bauxite containing 25–30% total iron. The metallic iron concentrate can be used as steelmaking material, and alumina can be extracted from the non-magnetic product. A general flow sheet of this process is shown in Fig. 7 [11]. This process can be suitably modified and adopted for metallurgical grade bauxite, to lower iron and titania contents, for value-added non-metallurgical applications.

7.2 Chemical Methods

A hydrochloric acid leaching study was carried out on Indian bauxite of Eastern Ghats (Odisha), to make the ore suitable for high-value, non-metallurgical applications [10]. The HCl leaching studies of bauxite indicated that as the concentration of HCl acid increases, the iron content reduces substantially, but at the same time, loss of alumina takes place. The change in bauxite colour from reddish brown to light colour indicated leaching of ore. It was found that bauxite of 300 μ size, leached with HCl of 25% concentration for 60 min at 95 °C, produced the best results. The recovery of high-grade bauxite (No. 1) was about 62%, and beneficiated bauxite (No. 2) produced at 40 min leaching had 76% recovery (Fig. 8). In the process, titania has also substantially decreased, with the total $TiO_2 + Fe_2O_3$ in the first beneficiated

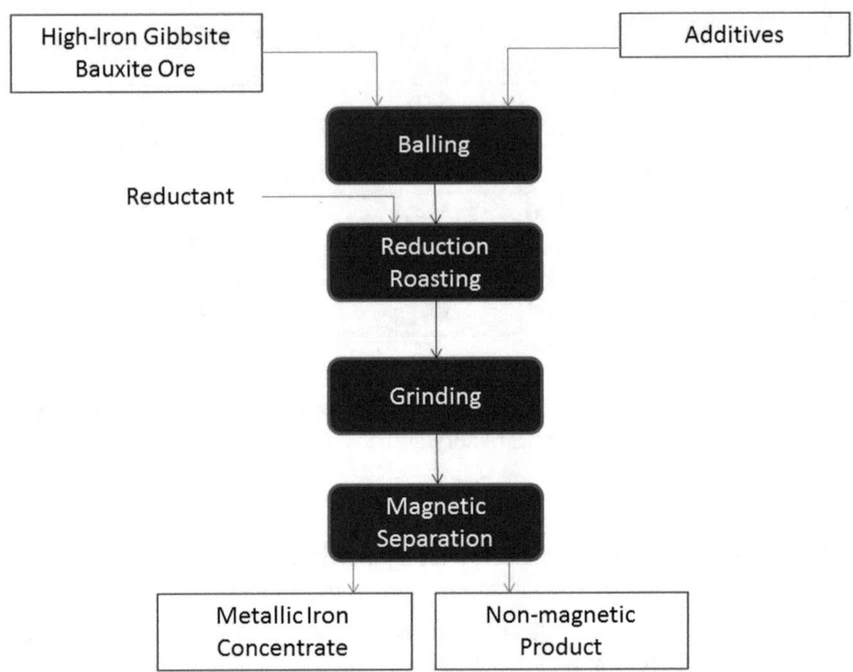

Fig. 7 A flowsheet of reduction roasting followed by magnetic separation

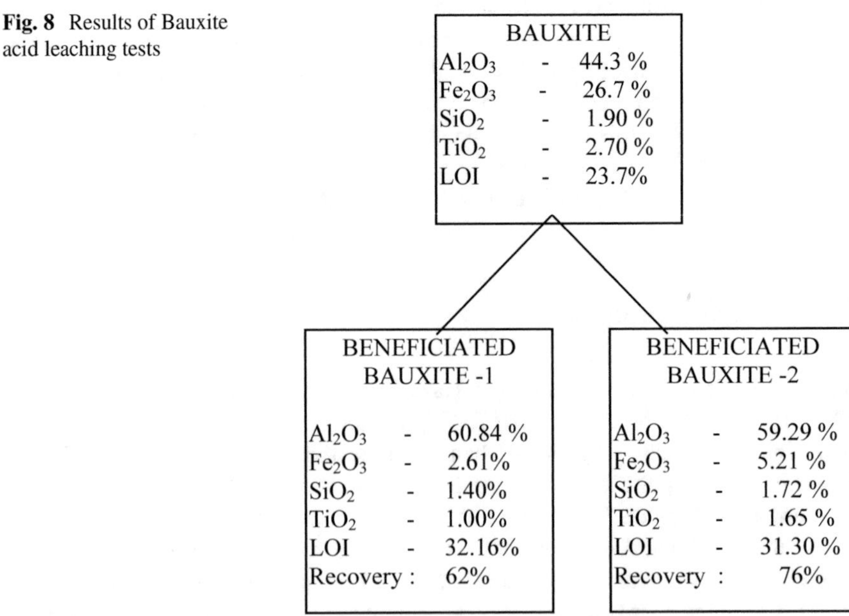

Fig. 8 Results of Bauxite acid leaching tests

Table 7 HCl leaching of santa catarina bauxite of Brazil

Main oxides	Natural bauxite (%Wt.)	Processed bauxite (%Wt.)
Al_2O_3	55.4	62.7
Fe_2O_3	5.8	0.6
SiO_2	3.5	5.5
TiO_2	0.9	0.6
ZrO_2	0.2	0.1
Loss on ignition	30.4	40.5

bauxite at only 3.61%, which can be considered as best quality of non-metallurgical grade bauxite.

It is preferred that before the acid leaching, the iron content of metallurgical bauxite can be brought down by physical beneficiation processes, such as the high intensity wet magnetic separation and reduction roasting followed by magnetic separation, so that minimum HCl can be used in bringing down Fe_2O_3 below 2.5%. On a similar concept, Altech Chemicals Ltd. has set up a high-purity alumina plant in Malaysia, where the standard HCl leach process is adopted to extract alumina and the raw material used is white clay.

A bauxite mining company of Brazil [12] carried out pilot tests for the acid leaching process in their mine, to produce one of the best quality low-iron bauxites for the refractory industry as shown in Table 7.

In order to minimize the input material for the leaching process, it is necessary to physically beneficiate bauxite by high-intensity wet magnetic separation and/or reduction roasting followed by magnetic separation, to lower the iron content. Acid leaching may be employed in special cases as this requires special reactor vessels, with necessary safety arrangements for industrial application.

8 Conclusion

- The prime objective of bauxite beneficiation for the metallurgical industry is to reduce the reactive silica content of the ore for lower caustic consumption, and higher alumina recovery. This is followed by reduction in iron content (mainly goethite), organics, boehmite etc., resulting in the overall improvement in quality and increase in available alumina.
- Dry beneficiation processes such as crushing to a particular size, followed by elimination of fines by screening, are used in upgrading the bauxite quality in various mining operations of India, Sierra Leone and Guinea.
- Wet beneficiation processes involving specific washing and classification are most effective for lowering the concentrations of kaolinite, goethite, fine quartz, titania and amorphous phases of lateritic bauxite.

- In Brazil, Sierra Leone and Vietnam, high silica run of mine (ROM) ores are effectively washed and scrubbed to produce high quality metallurgical grade bauxite.
- Thanks to the generally homogeneous nature of washed ore, combined with low kaolinite, goethite, organic carbon and high gibbsitic alumina content, these bauxites are the preferred ore for alumina refineries, dependent on import.
- Major non-metallurgical applications of bauxite include: high-temperature refractories, abrasives and ceramics industries, calcium aluminate cement, proppants, steel slag, and welding flux, etc. The specifications for these non-metallurgical bauxites tend to be more stringent compared to bauxites used in alumina plants.
- With the acute shortage of refractory and abrasive grade bauxites, it has become imperative to beneficiate widely available metallurgical grade bauxite, mainly by significantly lowering the iron and titania contents.
- There is still no proven commercial technology to produce high alumina, low iron and low titania bauxite suitable for refractory and abrasive industries, and industry is dependent on limited natural resources of Guyana, China and India. Here two research directions are proposed to produce value added non-metallurgical grade bauxite.

References

1. Kumar M, Senapati B, Satish Kumar C (2014) Beneficiation of high silica bauxite ores of india, In: 3rd Annual IBAAS conference, Visakhapatnam, November 2014, vol 3, pp 34–42. The IBAAS Binder
2. Datta B, Nandi A (2018) Bauxite beneficiation – issues & opportunities, In: 8th annual IBAAS conference, Mumbai, September 2018, vol 8, pp 35–48. The IBAAS Binder
3. Goswami R (2018) An initiative to remove silica from bauxite ore, In: 8th annual IBAAS conference, Mumbai, September 2018
4. Chaves AP (2010) Bauxite upgrading practices in Brazil. Travaux 35(39):110–116
5. Buntenbach S, Baumann T, Donhauser F (2010) Beneficiation of bauxite—upgrading of recoverable Al_2O_3. Travaux 35(39):117–124
6. AKW, Germany (2017) Unpublished report on pilot plant test of SMHL, Sierra Leone Bauxite (2017-15-15 BT Test Report + PSD 2016F96)
7. Urethane VorSpin Hydrocyclone (2020). http://www.cccmix.com/urethane-vorspin-hydrocyclone/. Accessed 25 Nov 2020
8. Tran A (2007) Quest for calcined bauxite, industrial minerals, pp 32–41
9. Nandi A (2017) Value addition and non-metallurgical uses of bauxite—opportunities in guinea. In: Annual IBAAS conference, Conakry, September 2017, vol 6, pp 67–75. The IBAAS Binder
10. Nandi A, Murthy PVR (1997) Production of special grade bauxites and export quality ore by de-ironing Eastern Ghats Bauxite of India. In: Proceedings of 8th international congress of ICSOBA, Milan, Italy, pp 16–18. Travaux 24(28):569–578
11. Li G, Sun N, Zeng J, Zhu Z, Jiang T (2010) Reduction roasting and Fe-Al separation of high iron content gibbsitic type Bauxite ores, Light Metals 133–137
12. Fernandes T, de Aquino H, Riella G, Bernardin AM (2011) Mineralogical and physical-chemical characterization of a bauxite ore from lages, Santa Catarina, Brazil, for refractory production. Miner Proc Extr Metall Rev Int J 32(3):137–149

Arsenic, from a Woeful Environmental Hazard to a Wishful Exploration Tool: A Case Study of Arsenic Contaminated Groundwater of Chakariya Area, Singrauli District, Madhya Pradesh, India

Gladson Bage, Tushar Meshram, Srinivasa Rao Baswani, M. L. Dora, Ajay Kumar Talwar, Abhinav Om Kinker, Manish Kumar Kewat, Hemraj Suryavanshi, Rajkumar Meshram, and Kirtikumar Randive

Abstract Arsenic is a toxic contaminant and an environmental health hazard. Regular consumption of arsenic beyond the permissible limit can cause systematic toxicity causing a severe threat to humans and animal life. Arsenic is a highly mobile element with a high response rate to even smaller fluctuations in temperature, pH, and redox conditions. However, it is not readily soluble in water in pure ionic form. The trivalent arsenic is readily absorbed by the skin, almost 60 times more readily than the pentavalent arsenic, absorbed by the gut. Under conditions of prolonged exposure, many organs may be damaged. The estimated geological contamination of groundwater has influenced a huge population world over. Though arsenic is highly toxic for life, it holds importance in economic geology by being a common pathfinder element for exploring gold and base metal sulfide (BMS) deposits. Systematic geochemical mapping of arsenic has proved to be an important tool in the exploration of Au and BMS deposits in Russia, China, and elsewhere. While investigating the arsenic environmental toxicity in the Chakariya area in the Singrauli district, MP, India, we found a systematic enrichment of arsenic near the

G. Bage
Geological Survey of India, Ranchi 834002, India

T. Meshram (✉) · S. R. Baswani · M. L. Dora · R. Meshram
Regional Petrology Division, Geological Survey of India, Nagpur 440006, India

R. Meshram
e-mail: rajkumar_meshram@rediffmail.com

A. K. Talwar · A. O. Kinker · M. K. Kewat
Geological Survey of India, Jabalpur 482003, India

H. Suryavanshi
NHM-II, Geological Survey of India, Nagpur 440006, India

K. Randive
Rashtrasanth Tukodoji Maharaj Nagpur University, Nagpur 440033, India
e-mail: randive101@yahoo.co.in

K. Randive et al. (eds.), *Innovations in Sustainable Mining*, Earth and Environmental Sciences Library, https://doi.org/10.1007/978-3-030-73796-2_8

mineralized quartz veins in the groundwater, soil, and stream sediment samples. The As analyses of water and stream sediments show values ranging from 1.2 to 27.5 mg/kg (or 1.2–27.5 ppm) having near-neutral pH (6.7–7.4), similar to a threshold value of 1–40 mg/kg (average 5 mg/kg) for soils, but much higher compared to water (0.01 mg/L). Arsenic can enter the water from a geogenic source or industrial and agricultural contaminants. Its concentration in the groundwater is high in areas of higher geothermal gradients such as geothermal fields. In the Chakariya area, the minerals arsenopyrite, scorodite, pyrite, and chalcopyrite emerged as the principal carriers of As in the mineralized zones in the surface and subsurface rock samples. Association of these minerals is often found with Au-Ag ± Pb-Zn-Cu ± Mo ± Cd ± Hg —mineralization. The Chakariya area, a part of the Mahakoshal Supracrustal belt (MSB), hosts several gold and sulfide mineralization pockets. However, any sizable deposit of Au-BMS remains lurking due to a lack of a systematic geoanalytical approach. The present study has demonstrated that arsenic, despite being a harmful environmental toxicant, provides a useful tool for tracking the hidden gold-BMS mineralization.

Keywords Arsenic contamination · Arsenopyrite · Scorodite · Mahakoshal supracrustal belt

1 Introduction

Groundwater is an essential natural resource for domestic and industrial water supply as well as agricultural purposes. Groundwater caters to ~50% of the current need for drinking water supplies of the world [1] and a source of about one-third of fresh-water withdrawals [2]. Arsenic, which is one of the most toxic element occurs in the environment [3–5], can be introduced in the environment either by natural processes through naturally occurring arsenic-rich minerals are desorbed and dissolved or by anthropogenic actions [6–8]. Bedrock aquifer provides a vital source of water for people living in rural areas of India [9], which is sometimes contaminated by toxic elements such as As and F, with occasional Cr, Cu, Fe, Mn, Ni, Pb, and Zn in the groundwater, causing deleterious health effects [10]. Arsenic, being the most common toxic element found as a contaminant in the groundwater, by its presence beyond permissible limits, becomes a major setback in providing safe drinking water to millions of people in Asia and worldwide [11]. Its intake causes the highest health risk and is responsible for the morbidity and mortality of many people around the world [12]. As per an estimate, a rural population exposed to unsafe As levels by drinking untreated groundwater in India, China, Myanmar, Pakistan, Vietnam, Nepal, and Cambodia have grown to over 100 million [13]. The fluctuations in the environment's physicochemical conditions are responsible for weathering, dissolution, precipitation, ion exchange, coupled with various biological processes that result in a change of the groundwater quality of the area [14, 15]. The arsenic concentration in groundwater is also governed by pH and redox conditions [16, 17].

Geologically, arsenic concentration is also correlated with the gold and BMS mineralization and often influence large areas of the gold-mining districts [18]. Although the ores (native or refractory gold, arsenopyrite, arsenian pyrite, chalcopyrite, sphalerite and pyrites) are important for the economy, they are also responsible for severe health hazards in the contagious areas. The oxidation of arsenopyrite results in scorodite formation, a primary cause of groundwater's geogenic source [19–21]. The World Health Organization [22] permissible limit for arsenic is 0.01 mg/l for drinking water and FAO (Food and Agriculture Organization) permissible limit is 0.10 mg/l for irrigation water [23, 24]. However, the concentrations of non-contaminated soils range from 0.1 to 10 mg/Kg [25].

The present study was carried out to assess the arsenic contamination in the groundwater of Chakariya area, Singrauli District, Madhya Pradesh, India. The study involved the collection of water samples from the dug wells. It bore wells, followed by their geochemical analyses, including measurement of minor and trace element concentration by ICP-MS. These results have indicated an alarmingly higher concentration of As in the shallow aquifers (dug wells) and deeper aquifers (bore wells). Supporting evidence has emerged from the study of soil and stream sediment samples collected in a specific grid pattern (Ref. https://www.gsi.gov.in), showing signs and systematic variation in the As concentration. The study indicates that such variation of As shows a positive correlation with the Au. This observation demonstrated its utility as a potential pathfinder element for exploring the gold and the base metal sulfides. This paper discusses the implications of the As-contamination in the groundwater of the Chakariya area. It simultaneously highlights arsenic's role as a reliable geochemical proxy in search of Au-BMS mineralization.

2 Geochemistry of Arsenic

Arsenic is primarily derived from the natural weathering of bed rocks. However, it is highly dependent on the geochemical conditions of the groundwater. Arsenic is widely diffused in nature and is concentrated in mineral deposits, particularly those containing sulfides and sulfosalts. It accompanies many elements in their deposits, including Cu, Ag, Au, Zn, Cd, Hg, U, Sn, Pb, P, Sb, Bi, S, Se, Te, Mo, W, Fe, Ni, Co, and Pt metals. The presence of As in groundwater is generally derived from the weathering of As bearing sulfide minerals, associated with hydrothermally altered zones and metasedimentary rocks [13, 26, 27]. Secondary altered minerals influence the mobilization of arsenic in groundwater. Under reducing conditions, arsenic is released from metal oxyhydroxides to groundwater [28, 29], while under oxidizing conditions, As is absorbed onto the mineral surfaces, particularly in clay minerals. The arsenic concentration in groundwater is governed by temperature, pH and redox conditions [16, 17].

Several natural and anthropogenic sources are deemed responsible for the As contamination in groundwater. Arsenic occurs as a major constituent of more than 200 minerals [30] and the sorption and dissolution of naturally occurring As bearing minerals and alluvial sediments result in high As concentration in groundwater in the deltas and the alluvial plains, even if the As concentration in the solid phase is not very high [31, 32]. Arsenic can be introduced in the environment either by natural processes such as during atmospheric emissions or when naturally occurring minerals rich in arsenic are desorbed and dissolved or by anthropogenic activities such as mining, combustion of fossil fuels, metal extraction, timber preservatives, etc. [7, 8]. The atmosphere occurs in the form of 'arsine,' which is a very toxic gaseous As-containing compound formed in a highly reducing environment [33–35]. The decrease in pH favors arsine formation. In anoxic conditions, arsine is liberated from marshy soil and swampy surfaces along with monomethyl-arsine, dimethyl-arsine, and trimethyl-arsine [33–36]. Oxidation leads to converting such gases to aqueous phases of As (V)-bearing compounds [33, 37, 38]. In the atmosphere, As is prominently present as particles. Volcanic eruptions, wind mobilization, marine aerosols, and industrial exhausts give rise to As in the atmosphere. These particles settle on the ground when fossil fuels are burnt or when smelters are used and are termed as wet or dry deposition [33, 39]. Wet deposition is that portion of atmospheric As, which is dissolved in rainwater [33].

Arsenic in groundwater exists primarily as oxyanions representing two oxidation states; arsenic As^{III} (arsenite) and arsenic As^{V} (arsenate) [3, 40]. Both As^{III} and As^{V} exist within the pH range of 6–9. The predominant As^{III} species are uncharged H_3AsO_3, while the primary arsenate species are monovalent H_2AsO^{-4} and divalent $HAsO_2^{-4}$. The geology and groundwater regime makes one of the forms (As^{III} or As^{V}) more dominant than the other [41, 42]. Although As^{V} is thermodynamically favored in oxic waters and As^{III} in anoxic waters, they have also been reported to coexist in both types of waters [43, 44]. Several studies have explained arsenic behavior and characteristics in the environment to understand its solubility and mobility [45, 46]. The toxicity and the removability of arsenic differ between As^{III} and As^{V}. The As^{III} is considered more toxic and more difficult to remove from water than the As^{V} [47]. The variability of the arsenic concentration in groundwater is ascribed to the aquifer's arsenic content and the varying dissolution/desorption processes releasing the arsenic from the solid phase to the liquid phase [41, 42, 48, 49]. Reductive dissolution of Fe oxides is considered the principal cause of release from aquifer sediments [50].

3 Geology and Geomorphology of the Area

The ENE-WSW trending Mahakoshal Supracrustal Belt is a prominent fault-bounded asymmetrical rift basin [51, 52] in Central India., The Mahakoshal Belt along the northern part of the Central Indian Tectonic Zone (CITZ) (Fig. 1a). It is bounded by Son-Narmada Northern Fault (SNNF) in the north and the Son-Narmada Southern Fault (SNSF) in the south and extends for more than 600 km along strike (Fig. 1b) [53, 54]. Metasediments dominate the Belt with subordinate mafic/ultramafic metavolcanics, including banded iron formation, dolomite, phyllite and quartzite, in the decreasing order of their abundance (Fig. 2). Stratigraphically, it is divided into

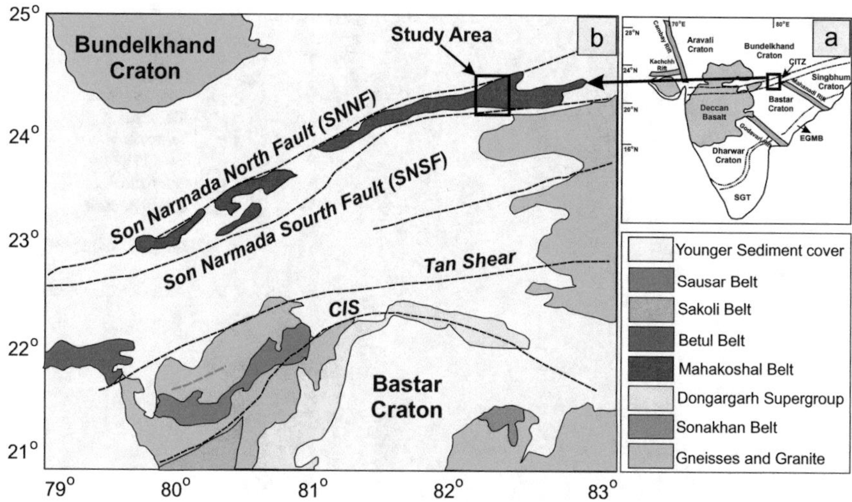

Fig. 1 **a** Inset map of India shows major cratons with rift basins within peninsular India. **b** Regional geological map shows the position of Mahakoshal Belt within Son Narmada North Fault (SNNF) and Son Narmada South Fault (SNSF) within Central Indian Tectonic Zone CITZ (modified after [54]

Parsoi and Agori Formations [53–56]. Both the formations are intruded by some quartz veins, at places mineralized and contain gold and BMS mineralization. The quartz veins of varying dimensions and colors (milky white, rosy pink and smoky black) are present in the Chakariya Block, targeted for gold and BMS mineralization for the last three decades [57]. The main source of gold in the Chakariya area is arsenopyrite and scorodite in the quartz veins associated with phyllite. The gold is generally refractory and occurs interlocked within structures of arsenopyrite and scorodite.

The study area topographically represents an undulating terrain having ridge and valley structure (Fig. 3a), and the elevation ranges from 360 to 400 m above the mean sea level (Fig. 1). The E-W flowing Son River is the main draining area and a significant drinking and irrigation source for the Singrauli District. A number of tributaries of the Son River are distributed in the Chakariya area, which follows dendritic and structurally controlled patterns. Some of these tributaries cross the Chakariya mineralized block, with scorodite exposures extending up to 700 m along the E-W direction (Fig. 3b–d).

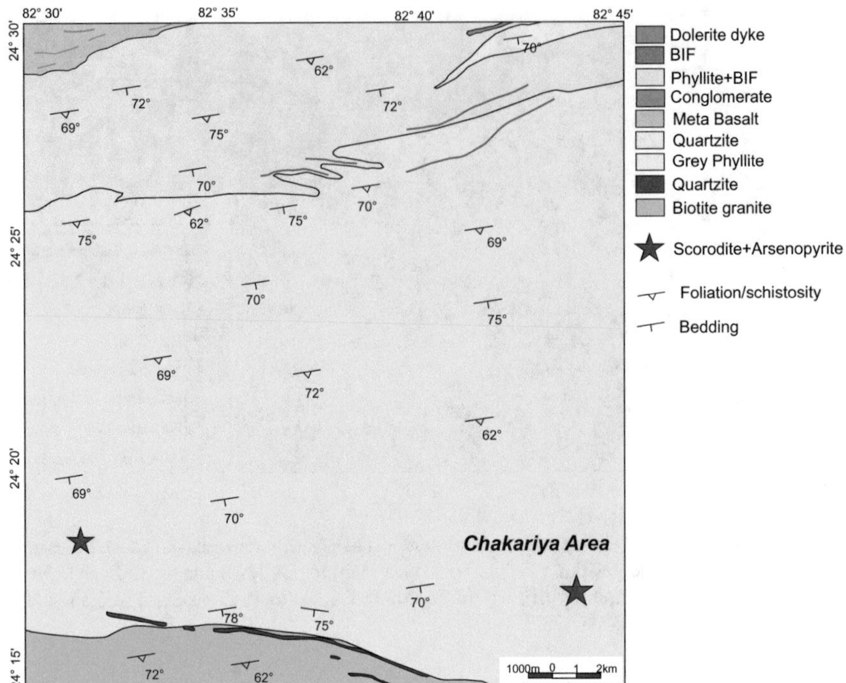

Fig. 2 Geological map of the study area (Chakariya) showing litho units belonging to Mahakoshal Group, Central India

4 Methodology

4.1 Sampling

A total of 14 water samples were collected from the dug wells and bore wells and running stream water from the Chakariya Block. The water samples were collected in clean polythene tarson bottles of 1 L capacity from each source. Soon after collecting water samples, the pH, temperature, and electrical conductivity (EC) were measured onsite. The water samples were then filtered using Whatman 0.45 μ filter paper and fixed using HNO_3 in the field and were taken to the laboratory for arsenic and other metals analyses. Similarly, the samples were collected from the scorodite — arsenopyrite-rich quartz vein for petrographic study and checked the host mineral phases' initial arsenic content.

4.2 Analytical Procedure

Analysis for CO_3^-, HCO_3^-, OH^-, Cl^-, F^-, NO_3^-, SO_4^-, T.H., Ca^{++}, Mg^{++}, Na^+, K^+, Li^+, TDS, SiO_2 for the water samples collected from the study area were carried out at the Geochemical Laboratory, Geological Survey of India, Central Region, Nagpur. ICP-MS directly analyzed the acid fixed filtered water samples.

The thin and polished sections (TPS) were at the Regional Petrology Laboratory, Geological Survey of India, Central Region, Nagpur and studied under the Leica DMRX Polarising Microscope. The scorodite and arsenopyrite phases were identified, and representative microphotographs were taken. The selected samples were analyzed using Electron Probe Micro Analyser (EPMA). Transmitted and reflected study of the samples was done. Mineral phases were also identified by CARL ZEISS-EVO 40 Scanning Electron Microscope coupled with Energy Dispersive Spectrometer at 20 kV and 63 eV detector resolutions at the Geological Survey of India (GSI), Hyderabad and Nagpur, India. Cameca SX-100 Electron Microprobe Analyser (EPMA) equipped with five WDS spectrometers at the Geological Survey of India laboratories at Hyderabad, India. The operating conditions were 20 kV accelerating voltage, 20 nA current intensity, and a beam diameter of 1 μm. Peak and background counting times for trace elements were 40 and 20 s, respectively. The representative analysis of water, soil and EPMA are presented in Table 1, respectively.

5 Results

5.1 Field Characteristics and Petrographic Description of Arsenic Bearing Sulfide Phases

The mineralized quartz veins are smoky in color and highly fractured in nature (Fig. 3b–d). The arsenopyrite, pyrite, chalcopyrite is the principal sulfide minerals observed in these mineralized zones on the surface and the drill cores (Fig. 3e–f). The arsenopyrite (Fig. 4a) and scorodite (hydrated iron arsenate, which is a product of supergene enrichment from the oxidation of arsenopyrite) occur in the form of fine disseminations, thin bands and stringers (Fig. 4b). Arsenopyrite occurs as euhedral to subhedral crystals with characteristic rhombic to angular shapes and shows metallic grey luster (Fig. 4a, c–e). At places, it occurs as fine-grained well-developed crystals aggregates (Fig. 4e). Arsenopyrite (few mms to 1 cm size) (Fig. 3e, f) by virtue of its silverfish white nature. The scorodite occurs as light yellowish green colored grains in the white or smoky quartz vein (Fig. 3b–d). The scorodite weathers to limonite, leaving reddish brown-yellowish ochre impregnations over the samples (Fig. 3c, f). The BSE images also show the arsenopyrite rim altered to scorodite (Fig. 4b) and pseudomorphs of arsenopyrite converted to iron oxide (Fig. 4f).

The compositional variation of arsenopyrite and scorodite is presented in Table 1. The arsenopyrite has higher As: 43.98–46.28%, Fe: 35.01–35.72%, S: 21.10–22.16%

Table 1 Representative EPMA analysis of arsenopyrite and Scorodite (wt%) from the Chakariya area, Mahakoshal Belt

Pt Analysis	Name	S	Fe	Au	Co	Ni	Zn	As	Ag	Te	Pb	Bi	Cu	Sb	Cd	Total
8/1	Arsenopyrite	21.106	35.012	0	0.013	0	0	45.217	0.466	0	0	0.052	0	0	0.008	101.874
14/1		21.294	35.224	0	0.013	0.024	0	46.288	0.108	0	0	0.006	0	0	0.028	102.984
15/1		22.106	35.369	0	0.03	0.02	0	44.676	0.123	0	0	0.183	0	0	0	102.507
16/1		21.828	35.729	0	0.042	0.005	0.01	44.936	0	0	0	0.087	0	0	0.034	102.672
17/1		22.165	35.137	0	0.04	0	0	43.987	0.139	0	0	0.152	0	0	0	101.621
18/1		22.17	35.512	0	0.046	0.033	0	44.579	0	0.006	0	0.049	0	0	0	102.396
19/1		21.35	35.192	0	0.039	0	0	45.382	0	0	0	0.068	0	0	0	102.031
23/1		21.727	35.598	0	0.067	0	0	45.476	0	0	0	0.024	0	0	0.02	102.912
24/1		21.74	35.496	0	0.015	0	0	44.848	0.122	0	0	0.261	0	0	0	102.481
25/1		21.826	35.708	0	0	0	0	44.488	0	0	0	0.04	0	0	0.044	102.107
9/1	Scorodite	0.433	21.026	0	0	0.019	0.181	36.698	0.33	0	0.128	0	0	0	0.843	59.658
10/1		0.231	17.713	0	0.018	0.045	0.24	38.102	0.054	0	0	0.083	0	0	0.104	56.59
11/1		0.377	20.865	0	0.008	0	0.178	37.234	0	0	0.081	0	0.094	0	0.593	59.429
12/1		0.495	20.839	0	0.012	0	0.191	36.843	0	0	0	0.089	0.015	0	0.9	59.384
13/1		0.351	21.062	0	0	0	0.285	36.824	0.044	0	0.048	0.013	0.086	0	0.721	59.433
20/1		0.348	20.659	0	0.034	0	0.273	38.117	0.158	0	0.081	0.004	0.037	0	0.671	60.383
21/1		0.535	20.69	0	0.021	0	0.276	37.494	0	0	0.034	0	0.049	0	0.921	60.019
22/1		0.426	20.916	0	0	0	0.185	37.059	0.095	0	0.061	0.038	0.11	0	0.695	59.585

Fig. 3 **a** Panoramic view of Chakariya area showing undulating terrain with small hillocks and depressions. Field photograph showing **b** development of scorodite within quartz vein. **c** ochre yellow sulfide stains on the surface of the mineralized quartz vein. **d** replacement of arsenopyrite into scorodite at Chakariya. **e** primary steel-grey arsenopyrite mineralization in the subsurface drill core sample (*scale as coin of Indian rupee two having a diameter of 2.5 cm*). **f** Hand specimen showing the development of scorodite and iron oxide from the arsenopyrite

with low content of Ag: 0.1–0.46%. Similarly, scorodite also has higher As: 36.69–38.11%, Fe: 17.71–21.06% with very low content of S: 0.23–0.53%, Zn: 0.17–0.28%, Ag: 0.04–0.33%, Pb: 0.03–0.12%, Bi: 0.004–0.08% and Cd: 0.1–0.92% (Table 1).

5.2 *Physicochemical Properties of Water Samples*

The selected physicochemical properties of groundwater samples from Chakariya block, Singrauli district, Madhya Pradesh are presented in Table 2. In the present study, it was observed that both stream and groundwater are neutral to alkaline. However, more alkalinity was observed in the stream water (pH = 6.7–7.4), perhaps due to the discharge of detergents in the stream during washing on the river banks. The

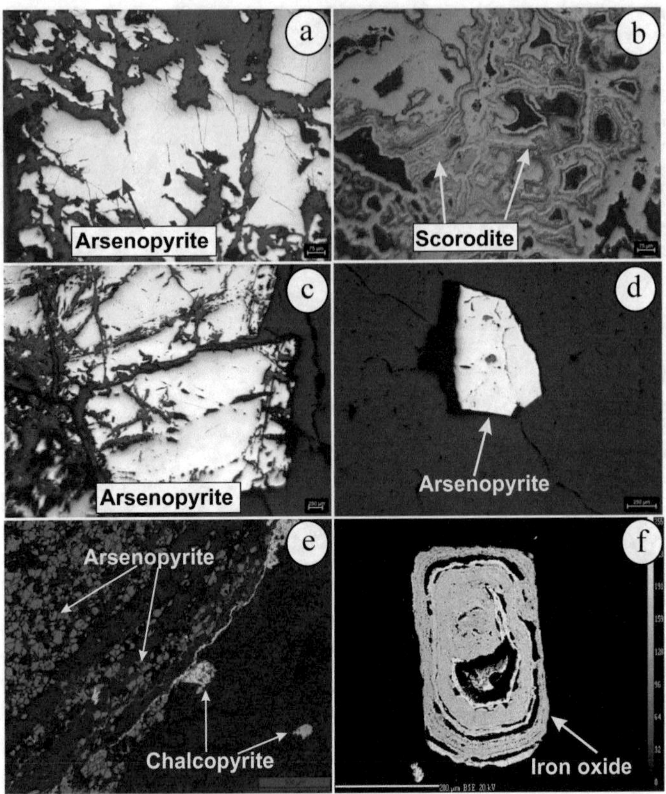

Fig. 4 Photomicrograph showing **a** metallic grey arsenopyrite under reflected light (oil immersion-20X). **b** arsenopyrite rim altered to scorodite under reflected light (oil immersion-20X), (**c–d**) highly fracture nature of arsenopyrite under reflected light (oil immersion-50X). **e** mineralized quartz veins with disseminated arsenopyrite and vein filling of chalcopyrite (reflected light-2.5X). **f** The BSE image showing pseudomorph of arsenopyrite converted to secondary iron oxide

electrical conductivity (EC) ranges from 380 μS/cm to 346 μS/cm and temperature varies from 21.62 °C to 27.92 °C. The Total Dissolved Solids (TDS) of groundwater and surface water ranges from 205 to 247 mg/L. The TDS is within a permissible range for drinking as well as agricultural purposes. Further, groundwater has the concentrations of Na vary from 23.7 to 44.6 mg/L (mean 33.76 mg/L), K from 0.4 to 1.8 mg/L (mean 0.76 mg/L), Ca from 14 to 22 m/L (mean 17.2 mg/L) and Mg from 4 to 18 mg/L (average 8.2 mg/L), respectively. Likewise, concentrations of Cl ranges from 0 to 84 mg/L (avg. 16.8 mg/L), NO_3 from 15.6 to 34.54 mg/L (avg. 6.90 mg/L), SO_4 from 09 to 40 mg/L (avg. 8 mg/L) and HCO_3 from 183 to 843 mg/L (avg. 169 mg/L) (Table 2).

A Piper diagram (Fig. 5a, b), which is widely used for determining sources of the dissolved constituents in water, has been prepared for surface and groundwater using an Easy_Quim v 5.0 (2012) software suggested the presence of a noticeable

Table 2 Physico-chemical properties of ground water samples from Chakariya block, Singrauli district, Madhya Pradesh

Sr. no	Parameter	W1	W2	W3	W4	W5
	Type of well	Dug well	Dug well	Dug well	Tube well	Dug well
1	Altitude (m)	335	336	343	347	348
2	Electric Conductivity (μS/cm)	346	342	380	391	412
3	pH	6.70	7.35	7.44	6.94	7.40
4	Temperature (°C)	24.57	21.62	21.63	27.92	25.70
5	TDS (mg/L)	208	205	228	235	247
6	pH	6.7	7.4	7.4	6.9	7.4
7	Conduct	346	342	380	391	412
8	TDS	208	205	220	235	247
9	HCO_3	165	165	177	183	153
10	CO_3	0	0	0	0	0
11	Chloride	0	0	0	0	84
12	SO_4	8.1	7.8	8.2	9	6.8
13	Ca	16	14	16	22	18
14	Mg	18	4	4	6	9
15	$CaCO_3$	115	50	55	80	80
16	Na	27.3	33.2	40	23.7	44.6
17	K	0.4	0.6	0.6	1.8	0.4
18	SiO_2	35	33.5	38	39	40.5
19	F	0.56	0.64	0.78	0.42	0.9
20	NO_3	5.1	8.2	15.6	5.3	0.34
21	PO_4	0.25	0.26	0.4	0.22	0.43
22	As (ppm)	7.5	3.63	27.49	1.21	5.13
23	Bi (ppm)	<0.02	<0.02	<0.02	<0.02	<0.02

Hardness as CaCO3, (All values in ppm except Conductivity in μs/cm)

concentration of Na, K and Ca rather than Mg, SO_4, CO_3, and HCO_3. The samples plots in low to medium hazardous field in the US salinity diagram (Fig. 5c, d). Similarly, the Gibbs diagram is also plotted using TDS Vs. Na/(Na + Ca) (Fig. 6a, b). The study area plots' water chemistry plots clearly in the rock interaction field (Fig. 6a, b), indicating a chemical process between water rock interaction includes dissolution, ion exchange, oxidation, and reaction play a vital role in liberating the As in the water.

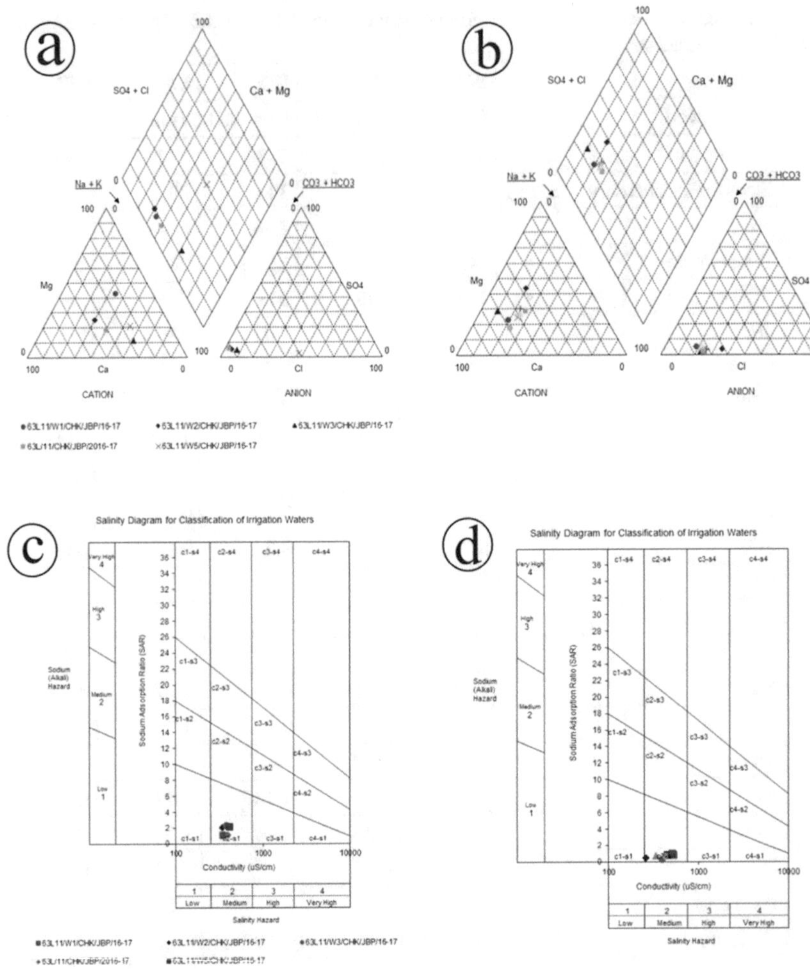

Fig. 5 **a** Piper diagram of surface water samples of Chakariya area. **b** Piper diagram of groundwater samples from Chakariya area. **c** The US salinity diagram for the classification of surface water in the Chakariya area. **d** The US salinity diagram for the classification of groundwater in the Chakariya area

5.3 Concentration of Arsenic in the Water Samples

The concentrations of arsenic in the water samples are given in Table 2. The arsenic range concentrations (0.5–3.3 mg/kg) are variable in the stream water, showing normal to the slightly higher range (Table 3). In contrast, the study area's groundwater has become highly contaminated with arsenic (1.2–27.5 mg/kg) (Table 2). The result shows that arsenic concentration in the stream water was above the permissible limit [23, 24]. The schematic diagram (Figs. 7 and 8) show how the arsenic source rock

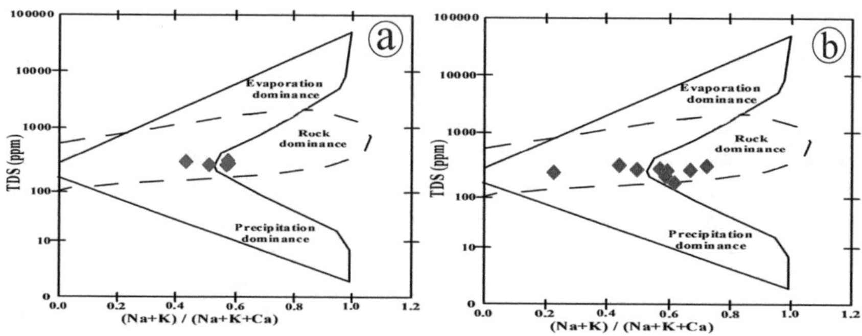

Fig. 6 The study of groundwater chemistry from the Chakariya block [58]

Table 3 In-situ measurement of physico-chemical paratmeters of stream water sample and arsenic values with associated stream sediments from Chakariya block, Singrauli district, Madhya Pradesh

Sr. no	Parameter	W1	W2	W3	W4	W5	W6	W7	W8	W9
1	Altitude (m)	282	289	327	291	323	339	409	342	244
2	Electric conductivity (μS/cm)	444	508	338	433	452	419	389	260	522
3	pH	7.40	8.01	*6.7*	8.14	7.65	*6.90*	*6.84*	7.06	7.56
4	Temperature (°C)	23.6	24.5	23.3	24.2	23.7	20.7	22.8	20.5	22.5
5	TDS (mg/L)	266	304	202	259	271	251	233	156	313
6	As[a] (ppm)	0.5	1.12	0.5	3.37	0.5	1.08	1.97	1.73	1.93
7	Bi [a] (ppm)	<0.02	<0.02	<0.02	<0.02	<0.02	<0.02	<0.02	<0.02	<0.02

[a]Stream sediment analysis

interaction with different natural weathering agencies leads to the disintegration, dissolution and mobilization of As in the area.

6 Discussion

6.1 *Weathering of Arsenopyrite and Release of Arsenic in Groundwater*

Almost all the metallic elements form complexes with oxygen and sulfur. Therefore, they occur either as oxides or sulfides depending upon relative oxidized or reduced environments in which they form or mobilize in the earth's crust. The common As-bearing naturally occurring compounds (minerals) are arsenopyrite, orpiment, realgar, and arsenian pyrite. As forms bonds; or chalcopyrite, pentlandite, pyrrhotite, sphalerite and pyrite, As is accommodated in crystal lattices by diadochy [59]. When these minerals are associated with gold mineralization, they show exceptional enrichment in the arsenic concentration. However, arsenopyrite has been the most common arsenic source released in water [60–63]. This mineral produces soluble arsenic in surface and groundwater resources by prolonged weathering, transportation, and long-term interaction with the groundwater. In the groundwater system, either recharge of freshwater or the drop in water-level caused by pumping could affect the oxidation of sulfide minerals under subsurface conditions. These fluctuating conditions increase arsenic dissolution during the reduction and formation of Fe-oxy hydroxides [64–66]. The arsenopyrite (FeAsS) first gets oxidized to scorodite ($FeAsO_4.2H_2O$) under low pH and presence of Fe^{3+}, following reaction (1) [19].

$$FeAsS + 14Fe^{3+} + 10H_2O \rightarrow 14Fe^{2+} + SO_4^{2-} + FeAsO_4 \cdot 2H_2O + 16H^+ \quad (1)$$

The scorodite gets dissolved in water to form arsenic oxyanions and iron hydroxide. However, this can occur either through incongruent ($pH > 4$) reaction (2) or congruent ($pH < 4$) reaction (3) [63, 67, 68].

$$FeAsO_4 \cdot 2H_2O + H_2O = H_2AsO_4^- + Fe(OH)_3(s) + H^+ \quad (2)$$

$$FeAsO_4 \cdot 2H_2O = H_2AsO_4^- + Fe(OH)_2^+ + OH^- \quad (3)$$

The study area's water samples invariably show pH values more than 4 (Tables 2 and 3), indicating that the incongruent reaction was dominant. A proton was released, which might have lowered pH at the local level. However, due to the high alkalinity of the waters (alkalinity 50–115 mh/L), the effect of proton-induced acidity might have got neutralized. Thus, the arsenic concentration of the waters in the Chakariya block is due to a geogenic source, i.e., an alteration of arsenopyrite from the host rocks (Figs. 3, 4 and 6). Consequently, the remobilization of As content from the source material can be caused by changes in environmental conditions due to fluctuation in temperature, pH and redox conditions. The released As can therefore be retained in the geochemical environment through either sorption onto, e.g., natural Fe(III) (oxy) hydroxides or precipitation (mainly as Fe(III) arsenate).

The arsenic liberated from the source gets added into the groundwater and they're by contaminating the groundwater [69]. Figure 7 illustrated the nature and path of arsenic contamination of groundwater and stream sediments in the Chakariya area. Arsenopyrite and scorodite are the principal carriers of geogenic arsenic content in the area. Both these arsenic bearing minerals are present in significant amounts in the study area. This may further have distributed through different geological agencies

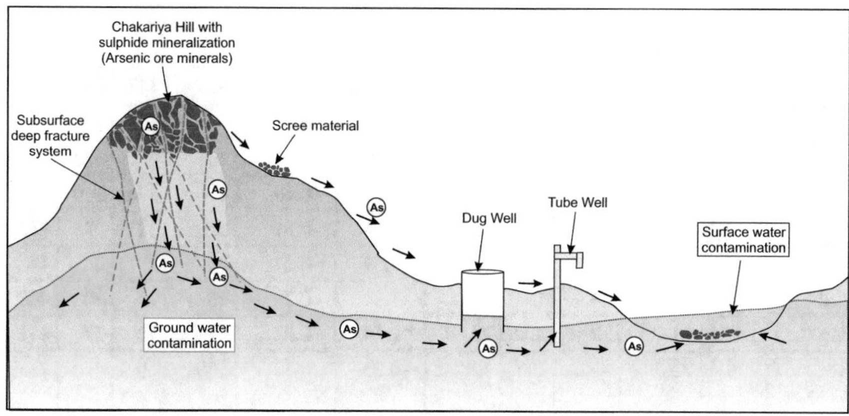

Fig. 7 Schematic diagram of arsenic contamination from source area to groundwater and surface water areas, Chakariya block, Madhya Pradesh

and deposited in the geomorphological low laying areas produced local or regional anomaly zones (Fig. 8).

6.2 *Geochemical Classification of the Groundwater*

Geology, water-rock interaction, and contamination sources constrain the quality of water-based on their chemical composition. Piper diagram of the studied samples is dominated in anion field by $HCO_3 + CO_3$ and Cation field shows slight enrichment in Mg and Ca (Fig. 5). Diamond-shaped quadrilateral field shows water as magnesium bicarbonate type (Fig. 5a, b). All the samples of the Chakariya area fall in low alkali hazard and medium salinity zone (Fig. 5c, d). To understand the genesis of water and the interaction between groundwater and aquifer minerals has a significant role in determining the water quality [70]. The ratio of Ca^{2+}/Mg^{2+} is used to determine sources of calcium and magnesium ions in groundwater [71]. A ratio of >2 indicates a substantial contribution of Ca^{+} and Mg^{+2} in the groundwater, which can be attributed to silicate minerals present in the host rocks [72].

The weathering of rocks is caused by the interaction between the water, rock, and earth's atmosphere (Figs. 6 and 7). The ratio of HCO_3/SiO_2 marks dominant weathering. The ratio of HCO_3/SiO_2 ration, which is <5, indicates the silicates' weathering [73]. All the groundwater samples of the Chakariya area show a ratio <5; whereas, the stream water samples show a ratio of >10, indicating enhanced anthropogenic activities (Tables 3 and 4).

According to the Gibbs diagram, the groundwater geochemistry from the Chakariya block falls in the rock-dominant zone indicating prolonged water-rock interaction. These samples suggest that rocks' chemical weathering was a dominant factor controlling groundwater chemistry of the study area (Fig. 7).

Table 4 Descriptive water statistics of water samples collected from Chakariya block, Singrauli district, Madhya Pradesh

	Mean	Median	Mode	SD	SV	Kurtosis	Skewness	Range	Min	Max	Sum
pH	7.37	7.4	#N/A	0.52	0.27	−1.56	0.11	1.4	6.7	8.1	66
E.C.	418.0	433	#N/A	81.5	6647	0.60	−0.78	262	260	522	3765
TDS	280.2	290	#N/A	54.6	299	0.60	−0.78	176	174	350	2522
Na^+	23.67	24	24	7.48	56.00	−0.64	−0.61	22	11	33	213
K^+	1.56	2	2	0.53	0.28	−2.57	−0.27	1	1	2	14
Ca^{++}	48.89	44	44	15.0	251	1.24	0.56	56	24	80	440
Mg^{++}	16.78	17	17	4.76	22.69	2.49	0.97	17	10	27	151
T.H as $CaCO_3$	236.6	220	#N/A	113.	12850.	−0.33	0.30	360	60	420	2130
SiO_2	11.67	10	9	4.12	17.00	6.15	2.38	13	9	22	105
HCO_3-	215	220	220	39.5	1562.0	1.54	−0.94	134	134	268	1940
CO_3--	5.33	0	0	10.5	112.0	0.73	1.62	24	0	24	48
NO_3--	4.37	4	#N/A	2.30	5.27	−0.77	0.56	6.5	1.8	8.3	39.3
Cl^-	38.33	38	43	5.15	26.50	−0.68	−0.67	14	29	43	345
SO_4--	8.22	8	9	3.23	10.44	−0.22	0.60	10	4	14	74
PO_4-	0.05	0.06	0.06	0.02	0.00	−1.11	−0.50	0.06	0.02	0.08	0.47

All values in ppm except pH and electrical conductivity (μS/cm)

Table 5 Classification of water quality based on suitability of water for irrigation purpose

Parameter	Range	Class	Samples (Dugwell/Handpump)	Samples (surface water)
EC	<250	Excellent	Nil	Nil
	250 to 750	Good	05	09
Na^+%	<20	Excellent	Nil	03
	20–40	Good	05	06
TH (Total Hardness)	<75	Soft	04	Nil
	75–150	Moderate	1	02
	150–300	Hard	–	05
	>300	Very Hard	–	02
MR (Magnesium Ratio)	<50	Suitable	05	09
	>50	Unsuitable	Nil	Nil

6.3 Potential Exploration Targets Using Arsenic as a Pathfinder Element

Indicator-, or pathfinder- elements are used to guide geochemical prospecting for various mineral deposits [74, 75]. They are represented by significant concentrations of a single or multiple elements (e.g., Ag, As, Bi, Sb, Cd, Cu, In, Mo, Pb, Te, W, Zn and REE) in the Earth due to supergene enrichment processes. Generally, these elements are restricted to a particular deposit because of specific intrinsic chemical properties. The Eh-pH conditions control the migration and concentration of elements, hydrolytic reactions, colloidal phenomena, and biological reactions, etc. [76, 77]. In particular, these indicator elements are generally dispersed and disintegrated in different media like bedrock, till, stream sediments, and stream water with associated ore deposits [78–84]. These are identified as primary or secondary dispersion halos and anomalous geochemical elemental distribution with threshold limit in the earth surfaces. Primary dispersion halos vary significantly in size and shape due to the various physical and chemical variables that affect rocks' fluid movements. These halos are detected at distances from few centmeter to hundreds of meters with respect to their ore bodies. Primary halos are controlling the following factors (i) presence or absence of fractures in the host rock, (ii) porosity and permeability of the host rock, (iii) fluid-rock interaction, and (iv) volatility of the ore elements (Table 5).

Indicator and pathfinder minerals are used in mineral exploration to narrow down the search area for many different ore deposits. When found as grains within sediments and rocks, these minerals give clues about the possible presence and

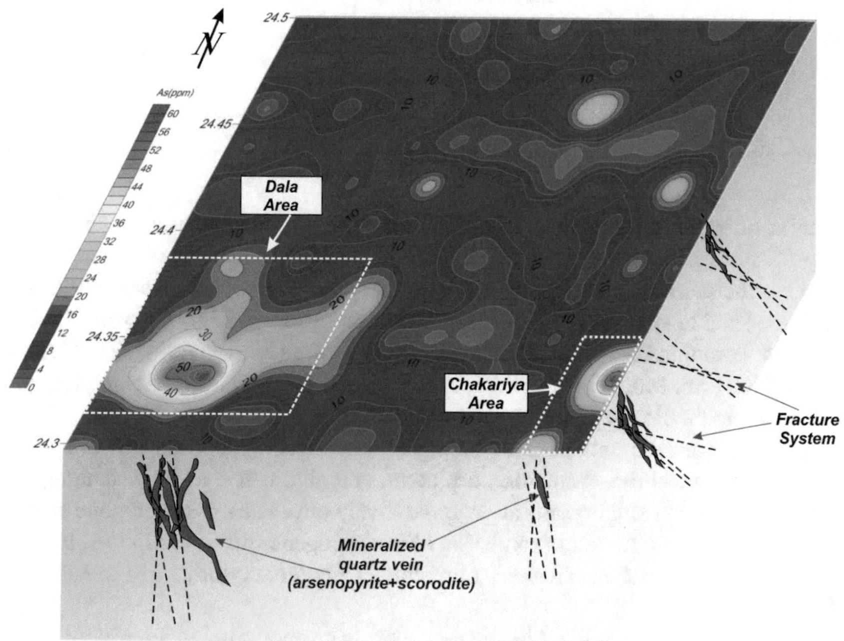

Fig. 8 The Elemental distribution map of Arsenic (As) in stream sediment samples was plotted using Surfer^{TM} software. The elemental distribution map of As shows a highly anomalous concentration (10–18 ppm) around Chakariya. The subsurface projection of the mineralized quartz vein is based on field observation and reported as anomalies

Table 6 Summary of pathfinder elements, minerals and associated ore deposits

S. no.	Pathfinder elements	Pathfinder mineral	Mineral deposits	References
1	As, Cu, Fe	Arsenopyrite and scorodite	Orogenic gold deposits	[90, 94]
2	Zn, Cu, Pb	Gahnite	VHMS and metamorphosed massive sulphide deposits	[95–97]
3	Cr, Mg, Ti	Chromium diopside, chromium spinels, magnesian ilmenite, and pyrope garnet	Diamond rich kimberlites	[98, 99]
4	Ni, Cr, Cu	Biotite	Ni-Cu-PGE deposits	[85, 90, 100, 101]
5	Be, Ce, La, Y, Zr, Th, U, La, Nd	Thorite, Zircon, Monazite, Xenotime	Zr, Y, and heavy rare earth elements (HREE)	[83]
6	Zr, Ca–Na, Ce, La, Mg, Fe	Porphyry indicator minerals (PIMS)-plagioclase, zircon and apatite, or magnetite; porphyry vectoring and fertility tools (PVFTS)-Epidote, chlorite and alunite	Porphyry copper deposits, VHMS	[85, 96, 101, 102]
			W-Mo deposits	[97]

location of specific mineralization types. They are commonly used in the search for gold, nickel, tungsten, diamond and platinum-group metals deposits (Table 6).

The principal sulfide ore minerals commonly contain traces of arsenic (As), antimony (Sb) and bismuth (Bi) due to their close relationship in many deposits [78, 79]. The most common elemental associates of arsenic are Cu, Ag, Au, U, Zn, Cd, Hg, Bi, Sn, Pb, P, Sb, Mo, W, Fe, Co, Ni, PGE, Se, and S. Therefore; arsenic is mainly used as a good indicator in many geochemical surveys [78–85].

The chief arsenic minerals in endogene (hypogene) deposits are arsenopyrite, cobaltite, niccolite, tennantite, orpiment, enargite, native arsenic, realgar, and proustite. The other sulpho-salts are comparatively rare. The main supergene arsenic minerals, formed as a result of oxidation of the hypogene sulfides, sulpho-salts, etc., are scorodite, beudantite, olivenite, mimetite, arsenolite, erythrite, and annabergite [78].

Arsenic is commonly fixed in sulfides such as arsenopyrite, loellingite or pyrite in many metal deposits. Generally, arsenopyrite is dominantly hosted arsenic in its lattice structure. However, sometimes the element is closely bound with Au mineralization [86–90] as well as form widespread geochemical haloes around Au deposits. Hence arsenic is commonly used as a pathfinder in noble metal exploration [78,

86, 91–93]. The present study area is well known for its gold occurrences in the Mahakoshal belt. The gold mineralization is associated with some quartz veins that are intruded in the host phyllites. The gold is refractory and dominantly associated with arsenopyrite and scorodite.

Similarly, the area has undulating terrain with hillocks and depression with structurally control geomorphology and drainage pattern. The present study reported arsenic anomalies from stream sediments and groundwater samples supported their geogenic source. The arsenopyrite and scorodite are directly used as pathfinder/indicator minerals in the area to locate gold mineralization in the Mahakoshal belt. Similarly, the reported higher values of arsenic from two locations in the Chakariya area are supported by primary and secondary haloes in the area (Fig. 8). This further indicates that one zone represents a new location (near Dala village) with As supergene enrichments dispersion haloes out of two anomalies zones. It favors gold mineralization in that locality other than Chakariya block, which is shown by the schematic diagram, represented the anomalies zone and its subsurface connection. Finally, the present study also provides substantial proof that arsenic as a pathfinder element, is used to target potential exploration sites/areas.

7 Conclusion

The Mahakoshal Belt is well known for gold mineralization, associated with vein quartz in the phyllitic rocks. The present study reports a high concentration of arsenic in the groundwater and stream sediments in the Chakariya area, Singrauli district, MP as analyses of the water and stream sediments from the study area indicate that the values are ranging from 1.2 to 27.5 mg/kg (or 1.2–27.5 ppm) having near-neutral pH (6.7–7.4). The arsenic content in the water is higher than the permissible limit for drinking and irrigation. Hence, it is hazardous and toxic to human as well as animal life for regular consumption.

In contrast, these higher anomalies of arsenic, especially from stream sediments used as a very good pathfinder for gold mineralization in the area. The reported arsenic values in stream sediment and arsenopyrite and scorodite provide the clue about the occurrence of gold deposits in the area. These high arsenic values are due to the occurrence of arsenopyrite and scorodite and the formation of subsequent primary and secondary dispersion haloes over the source, which serve as a good indicator and play a vital role in locating gold deposits in the area. The present study brings to light two significant arsenic anomaly zones in Proterozoic Mahakoshal Belt. One is an over known Chakariya gold prospect and another area form located to the west of Chakariya around Dala village in Madhya Pradesh. This approach, based on using arsenic as a pathfinder element, will further help to locate such new gold mineralization occurrences by a similar geochemical approach in the Mahakoshal belt and other areas in Central India. Finally, it can be concluded that the arsenic is though an act as a woeful environmental hazard is also a wishful exploration tool.

Acknowledgements Authors extend sincere thanks to Dr. Ranjit Rath, Director General, and Shri. Janardan Prasad, Addl. Director-General and HoD, Geological Survey of India, Nagpur for granting permission to publish this paper. TM, thank Shri. Sanjeev Raghav, DDG, and RMH-III for his continuous encouragement to write this quality manuscript and Dr. Manash Roy Choudhry, Director and Mrs. Khipra Meshram, Sr. Geologist for their constant support during the processing of this manuscript. The Geological Survey of India entirely supported this work under different FSP. Due acknowledgment has been given to the Geological Survey of India for providing all facilities to conduct this research during the assigned Field Season Program (FS 2019–2020). The authors also acknowledge GSIs in house reviewers, and anonymous reviewers for their constructive and thoughtful comments, which helped in improving the quality of the manuscript.

References

1. Zektser IS, Everett LG (2004) Groundwater resources of the world and their use, p 346
2. Bondu R, Cloutier V, Rosa E et al (2018) Occurrence of geogenic contaminants in private wells from a crystalline bedrock aquifer in western Quebec, Canada: geochemical sources and health risks. J Hydrol 627–637. https://doi.org/10.1016/j.jhydrol.2018.02.042
3. Cullen WR, Reimer KJ (1989) Arsenic speciation in the environment. Chem Rev 89:713–764. https://doi.org/10.1021/cr00094a002
4. Dermatas D, Moon DH, Menounou N, Meng X, Hires R et al (2004) An evaluation of arsenic release from monolithic solids using a modified semi-dynamic leaching test. J Hazard Mater 116:25–38. https://doi.org/10.1016/j.jhazmat.2004.04.023
5. Hudson-Edwards KA, Houghton SL, Osborn A et al (2004) Extraction and analysis of arsenic in soils and sediments. Trends Anal Chem 23:745–752. https://doi.org/10.1016/j.trac.2004.07.010
6. Bissen M, Frimmel FH (2003) Arsenic: a review. Part I: occurrence, toxicity, speciation, mobility. Acta Hydrochim Hydrobiol 31:9–18
7. Bhattacharya P, Chatterjee D, Jacks G et al (1997) Occurrence of arsenic contaminated groundwater in alluvial aquifers from delta plains, eastern India: options for safe drinking water supply. J Water Res Develop 13:79–92
8. Bhumbla DK, Keefer RF (1994) Arsenic mobilization and bioavailability in soils. In: Nriagu JO (ed) Arsenic in the environment, part 1 cycling and characterization. Wiley, New York, pp 51–82
9. Perrin J, Ahmed S, Hunkeler D et al (2011) The effects of geological heterogeneities and piezometric fluctuations on groundwater flow and chemistry in a hard-rock aquifer, southern India. Hydrogeol J 19(6):1189–1201
10. Abiye TA, Mkansi S, Masindi K, Leshomo J et al (2018) Effectiveness of wetlands in retaining metals from mine water South Africa. J Water Environ 32:259–266
11. The World Bank (2005) Arsenic contamination of groundwater in South and East Asian Countries, vol I, Policy Report. (Washington, DC: Office of the Publisher, The World Bank)
12. Hopenhayn C (2006) Arsenic in drinking water: impact on human health. Elements 2:103–107
13. Ravenscroft P, Brammer H, Richards K et al (2009) Arsenic pollution: a global synthesis. Wiley-Blackwell, West Sussex, UK, p 618. ISBN: 978-1-405-18601-8
14. Todd DK (1980) Groundwater hydrology. Wiley, New York
15. Sakram G, Sundaraiah R, Saxena PR, Bhoopathi V (2013) The impact of agricultural activity on the chemical quality of groundwater, Karanjavagu watershed, Medak district, Andhra Pradesh. Int J Adv Scient Techn Res 3(6):769–786
16. Dixit S, Hering JG (2003) Comparison of arsenic (V) and arsenic (III) sorption onto iron oxide minerals: implications for arsenic mobility. Environ Sci Technol 37(18):4182–4189

17. Hug SJ, Leupin O (2003) Iron-catalyzed oxidation of arsenic(III) by oxygen and by hydrogen peroxide: pH-dependent formation of oxidants in the Fenton reaction. Environ Sci Technol 37(12):2734–2742. https://doi.org/10.1021/es026208x
18. Abiye TA, Bhattacharya P (2019) Arsenic concentration in groundwater: archetypal study from South Africa. Groundwater Sustain Develop 9. http://dx.doi.org/10.1016/j.gsd.2019.100246
19. Dove PM, Rimstidt JD (1985) The solubility and stability of scorodite, $FeAsO_42H_2O$. Amer Miner 70:838–844
20. Bhattacharya P, Jacks G, Frisbie SH et al (2002) Arsenic in the environment: a global perspective. In: Sarker B (ed) Handbook of heavy metals in the environment. (New York, NY: Marcel Dekker Inc.), pp 147–215
21. Nriagu J, Bhattacharya P, Mukherjee A, Bundschuh J et al (2007) Arsenic in soil and groundwater: an overview. In: Bhattacharya P, Mukherjee A, Bundschuh J, Zevenhoven R, Loeppert R (eds) Arsenic in soil and groundwater environment. Elsevier, Amsterdam, The Netherlands, pp 3–60
22. WHO (2007) Joint FAO/WHO Expert standards program codex alimentation commission. Geneva. http://www.who.int. Accessed 10 Sept 2012
23. Bhattacharya P, Samal AC, Majumdar J, Santra SC et al (2009) Transfer of arsenic from groundwater and paddy soil to rice plant (Oryza sativa L.): a micro level study in West Bengal India. World J Agricul Sci 5:425–431
24. Ahsan DA, Del Valls TA (2011) Impact of arsenic contaminated irrigation water in food chain: an overview from Bangladesh. Int J Environ Res 5:627–638
25. Kabata-Pendias A, Pendias H (1992) trace elements in soil and plants, 2nd edn. CRC, Boca Raton, FL
26. Bondu R, Cloutier V, Benzaazoua M et al (2017) The role of sulfide minerals in the genesis of groundwater with elevated geogenic arsenic in bedrock aquifers from western Quebec Canada. Chem Geol 474:33–44
27. Ryan PC, West DP, Hattori K et al (2015) The influence of metamorphic grade on arsenic in metasedimentary bedrock aquifers: a case study from Western New England, USA. Sci Total Environ 505:1320–1330
28. Lipfert G, Reeve AS, Sidle WC et al (2006) Geochemical patterns of arsenic-enriched ground water in fractured, crystalline bedrock, Northport, Maine, USA. Appl Geochem 21:528–545
29. Yang Q, Culbertson CW, Nielsen MG et al (2015) Flow and sorption controls of groundwater arsenic in individual boreholes from bedrock aquifers in central Maine, USA. Sci Total Environ 505:1291–1307
30. Chakraborti D, Mukherjee S, Pati S, Sengupta MK et al (2003) Arsenic ground water contamination in middle Ganga Plain, Bihar, India: a future danger. Environ Health Perspect 111:194–201
31. Chakraborti D, Rahman MM, Chowdhury UK et al (2002) Arsenic calamity in the Indian subcontinent: what lessons have been learned? Talanta 58:3–22
32. Chakraborty S, Nath B, Chatterjee D et al (2014) Retardation of arsenic transport by oxidized Holocene aquifer sediments of West Bengal India. J Hydrol 518:460–463
33. Kossoff D, Hudson-Edwards KA (2012) Arsenic in the environment. Chapter 1. In: Santini JM, Ward SM (eds) The metabolism of arsenite, arsenic in the environment, vol 5. London, CRC Press, pp 1–23
34. Chauhan S, D'Cruz R, Fauqi S, Singh KK et al (2008) Chemical warfare agents. Environ Toxicol Pharmacol 26:113–122
35. Sharma VK, Sohn M (2009) Aquatic arsenic: toxicity, speciation, transformations, and remediation. Environ Int 35:743–759
36. Duker AA, Carranza EJM, Hale M et al (2005) Arsenic geochemistry and health. Environ Int 31:631–641
37. Turpeinen RR, Panstar-Kallio M, Haggblom M et al (2002) Role of microbes in controlling the speciation of arsenic and production of arsines in contaminated soils. Sci Total Environ 285:133–145. https://doi.org/10.1016/S00489697(01)00903-2

38. Turpeinen R, Pantsar-Kallio M, Kairesalo T et al (2001) Role of microbes in controlling the speciation of arsenic and production of arsines in contaminated soils. Sci Total Environ 285:133–145
39. Plant JA, Kinniburgh DG, Smedley PL (2007) Arsenic Selen Treatise Geochem Elsevier 9(02):17–66
40. Ferguson F, Gavis J (1972) A review of the arsenic cycle in natural waters. Water Res 6(11):1259–1274
41. Welch AH, Westjohn DB, Helsel DR et al (2000) Arsenic in ground water of the United States: occurrence and geochemistry. Ground Water 38(4):589–604
42. Welch AH, Lico MS, Hughes JL et al (1988) Arsenic in ground water of the Western United States. Groundwater 26(3):333–347
43. Mok WM, Wai CM (1990) Distribution and mobilization of arsenic and antimony species in the Coeur D'Alene river Idaho. Environ Sci Technol 24(1):102–108
44. Andersen LCD, Bruland KW (1991) Biogeochemistry of arsenic in natural waters: the importance of methylated species. Environ Sci Technol 25(3):420–427
45. Haque SE, Johannesson KH (2006) Concentrations and speciation of arsenic along a groundwater flow-path in the Upper Floridan aquifer, Florida USA. Environ Geol 50(2):219–228
46. Kim MJ, Nriagu J, Haack S et al (2003) Arsenic behavior in newly drilled wells. Chemosphere 52(3):623–633
47. USEPA (2001) National primary drinking water regulations: arsenic and clarifications to compliance and new source contaminants monitoring. Federal Register CFR Parts 40(9):141–142
48. Hering JG, Kneebone PE (2001) Biogeochemical controls on arsenic occurrence and mobility in water supplies. In: Environmental chemistry of arsenic, pp 155–182
49. Jain CK, Ali I (2000) Arsenic: occurrence, toxicity and speciation techniques. Water Res 34(17):4304–4312
50. Guo H, Zhang B, Li Y et al (2011) Hydrogeological and biogeochemical constrains of arsenic mobilization in shallow aquifers from the Hetao basin, Inner Mongolia. Environ Pollut 159(4):876–883
51. Roy A, Bandyopadhyay BK (1990) Tectonic and strucutral pattern of of the Mahakoshal belt of Central India. A dicussion. In: Precambrian of Ce ntral India. Geol Surv India Spec Pub 288–240
52. GSI bulletin Series (2013) Gold mineralisation in son-valley gold belt. Parts of Sidhi and Sonbhadra, District, Madhya Pradesh and Uttar Pradesh, GSI Bulletin Series-A, p 61
53. Roy A, Devarajan MK (2000) Appraisal of the stratigraphy and tectonics of proterozoic mahakoshal supracrustal belt, central india. Geol Surv India Spec Publ 57:79–97
54. Roy A, Hanuma Prasad M, Devarajan MK et al (2002) Ductile shearing and synkinematic granite emplacement along the southern margin of Mahakoshal supracrustal belt: evidence from Singrauli area, Madhya Pradesh. J Geol Soc India 59:9–21
55. Mathur SM, Narain K (1981) Geosynclinal sedimentation in the Archaean of the Mirzapur-Sidhi area, in central India. Geol Surv India Spec Publ 3:31–37
56. Khanna T, Rao DV, Sai V, Satyanarayanan M et al (2020) ca. 2.1 Ga Mahakoshal supracrustal belt: an allochthonous terrain in central india tectonic zone. Lithos, pp 374–375. https://doi.org/10.1016/j.lithos.2020.105705
57. Bage G, Klinker AO (2016) Final report on general exploration for gold mineralisation in chakariya block, Tehsil-Chitrangi, Sidhi- District, Madhya Pradesh. Stage (G-2), Unpub. Report. https://www.gsi.gov.in
58. Gibbs RJ (1970) Mechanisms controlling world water chemistry. Science 170(3962):795–840
59. Randive SD (2012) Cultivation and study of growth of oyster mushroom on different agricultural waste substrate and its nutrient analysis. Adv Appl Sci Res 3(4):1938–1949
60. Buckley AN, Walker W (1988) The surface composition of arsenopyrite exposed to oxidizing environments. Appl Surf Sci 35:227–240. https://doi.org/10.1016/0169-4332(88)90052-9
61. Nesbitt HW, Muir IJ, Pratt AR (1995) Oxidation of arsenopyrite by air and air-saturated, distilled water, and implications for mechanism of oxidation. Geochim Cosmochim Acta 59:1773–1786. https://doi.org/10.1016/0016-7037(95)00081-A

62. Smedley PL, Kinniburgh DG (2002) A review of the source, behaviour and distribution of arsenic in natural waters. Appl Geochem 17:517–568. https://doi.org/10.1016/S0883-2927(02)00018-5
63. Basu A, Schreiber EM (2013) Arsenic release from arsenopyrite weathering: insights from sequential extraction and microscopic studies. J Hazard Mater 262:896–904
64. Matisoff G, Khourey CJ, Hall JF et al (1982) The nature and source of arsenic in northeastern Ohio ground water. Ground Water 20:446–456
65. Belzile N, Tessier A (1990) Interactions between arsenic and iron oxyhydroxides in lacustrine sediments. Geochim Cosmochim Acta 54:103–109
66. Deutsch WJ (1998) Groundwater geochemistry. Fundamentals and applications to contamination. CRC Press, 2000 Corporate Blvd., N.W. Boca Raton, FL 33431 (USA), p 232
67. Zhu Y, Merkel BJ (2001) The dissolution and solubility of scorodite, $FeAsO_4.2H_2O$: evaluation and simulation with PHREEQC2. In: Wiss. Mitt. Inst. für Geologie, TU Bergakedemie Freiberg, Germany, p 72
68. Harvey MC, Schreiber ME, Rimstidt JD, Griffith M (2006) Scorodite dissolution kinetics: implications for arsenic release. Environ Sci Technol 40:6709–6714
69. Bader M, Hartwig A, MAK Commission (2017) Arsenic and its inorganic compounds (with the exception of arsine) [MAK Value Documentation, 2015], 2, 4. https://doi.org/10.1002/3527600418.mb744038vere5817
70. Cedarstorm DJ (1946) Genesis of groundwater in coastal plains of Virginia. Environ Geol 41:218–245
71. Maya AL, Loucks MD (1995) Solute and isotope geochemistry and groundwater flow in the central Wasatch range Utah. J Hydrol 172:31–59
72. Katz BG, Coplen TB, Bullen TD, Davis JH et al (1998) Use of Chemical and isotopic tracers to characterise the interaction between groundwater and surface water in mantled karst groundwater 35:1014–1028
73. Hounslow AW (1995) Water quality data: analysis and interpretation. CRC Lewis publisher, Boca Raton, FL, pp 86–87
74. Bateman AM (1950) Economic mineral deposits, 2d edn. John Wiley and Sons Inc, New York
75. Warren HV, Delauvault RE (1959) Pathfinding elements in geochemical prospecting, in Symposium de Exploration Geoquimica. Int Geol Congr Twentieth Session Mexico 2:255–260
76. Boyle RW (1974) Elemental associations in mineral deposits and indicator elements of interest in geochemical prospecting (rev.), Canada. Geol Surv Paper 74–85
77. Boyle RW (1988) Indicator elements. In: General geology. Encyclopedia of earth science. Springer, Boston, MA
78. Boyle RW, Jonasson IR (1973) The geochemistry of arsenic and its use as an indicator element in geochemical prospecting. J Geochem Explor 2:251–296
79. Hale M (1981) Pathfinder applications of arsenic, antimony and bismuth in geochemical exploration. Develop Econ Geol 15:307–323
80. Bajc AF (1994) Gold grains in surface till samples, Parkin and Norman Townships, Sudbury. Ontario Geological Survey, Open File Report 5893
81. Averill SA (2013) Discovery and delineation of the Rainy River gold deposit using glacially dispersed gold grains sampled by deep overburden drilling. In: Paulen RC, McClenaghan MB (eds) New frontiers for exploration in glaciated terrain. Geological Survey of Canada, Open File 7374, pp 37–46
82. McClenaghan MB (2005) Indicator mineral methods in mineral exploration. geochemistry: exploration. Environ Anal 5:233–245
83. McClenaghan MB, Paulen R, Kjarsgaard I et al (2017) Rare earth element indicator minerals: an example from the Strange Lake deposit, Quebec and Labrador, eastern Canada. https://doi.org/10.4095/306305
84. McClenaghan MB, Ames DE, Cabri LJ et al (2019) Indicator mineral and till geochemical signatures of the Broken Hammer Cu-Ni PGE-Au deposit North Range, Sudbury structure, Ontario, Canada, Geological Society of London, https://doi.org/10.1144/geochem2019-058

85. McClenaghan MB, Paulen R, Oviatt N et al (2018) Geometry of indicator mineral and till geochemistry dispersal fans from the Pine Point Mississippi Valley-type Pb-Zn district, Northwest Territories, Canada. J Geochem Explor 190. https://doi.org/10.1016/j.gexplo.2018.02.004
86. Boyle RW (1979) The geochemistry of gold and its deposits (together with a chapter on geochemical prospecting for the elements), Canada. Geol Surv Bull 280:584p
87. Marion P, Holliger P, Boiron M et al (1991) New improvements in the characterization of refractory gold in pyrites: An electron microprobe, Mössbauer spectrometry and ion microprobe study. Brazil Gold 91:389–395
88. Arehart GB, Chryssoulis SE, Kesler SE et al (1993) Gold and arsenic in iron sulfides from sediment-hosted disseminated gold deposits; implications for depositional processes. Econ Geol 88:171–185
89. Friedl J (1993) Untersuchungen an Goldmineralen und goldhaltigen Erzen mittels Mössbauer-spektroskopie. Thesis, Technical University of Munich, Munich, Germany, p 149
90. McClenaghan MB, Cabri L (2011) Review of gold and platinum group element (PGE) indicator minerals methods for surficial sediment sampling. Geochem: Explor Environ, Anal 11:251–263. https://doi.org/10.1144/1467-7873/10-IM-026
91. Cavender WS (1963) Arsenic in geochemical gold exploration. Min Eng 15:60
92. Zhu B, Yu H (1995) The use of geochemical indicator elements in the exploration for hot water sources within geothermal fields. J Geochem Explor 55(1–3):125–136
93. Akçay M, Özkan HM, Moon CJ et al (1996) Secondary dispersion from gold deposits in west Turkey. J Geochem Explor 56:197–218
94. Goldfarb RJ, Leach DI, Pickthorn WA et al (1988) Origin of lode gold deposits of the Juneau gold deposit, southeast Alaska. Geology 16:440–443
95. Spry P, Teale G (2009) Gahnite composition as a guide in the search for metamorphosed massive sulfide deposits. Int Assoc Appl Geochem Indicat Miner Workshop B Freder 27–34
96. McClenaghan MB, Peter J, Layton-Matthews D et al (2015) Overview of VMS exploration in glaciated terrain using indicator minerals, till geochemistry, and boulder tracing: a Canadian perspective
97. McClenaghan MB, Parkhill MA, Pronk AG et al (2016) Indicator mineral and till geochemical signatures of the mount pleasant W-Mo-Bi and Sn-Zn-In deposits, New Brunswick, Canada. J Geochem Explor 172. https://doi.org/10.1016/j.gexplo.2016.10.004
98. Woolley AR, Bergman SC, Edgar AD et al (1996) Classification of lamprophyres, lamproites, kimberlites, and the kalsilitic, melilitic, and leucitic rocks. Can Miner 34:175–186
99. McClenaghan MB, Kjarsgaard B (2007) Indicator mineral and surficial geochemical exploration methods for kimberlite in glaciated terrain; Examples from Canada. Miner Dep Canada Syn Major Deposit Types District Metall Evol Geol Prov Explor Methods 5:983–1006
100. Averill SA (2011) Viable indicator minerals in surficial sediments for two major base metal deposit types: Ni-Cu-PGE and porphyry Cu. Geochem: Explor Environ, Anal 11:279–291
101. David J, Robert A, Mark B et al (2010) Porphyry copper deposit model
102. Bower B, Payne J, DeLong C et al (1995) The oxide-gold, supergene and hypogene zones at the Casino gold-copper-molybdenum deposit, west-central Yukon. In: Schroeter TG (ed) Porphyry deposits of the northwestern Cordillera of North America: Canadian Institute of Mining, metallurgy and petroleum special, vol 46, pp 52–366

Recovery of Valuable Metals (Rare-Earths) from Phosphate Fertiliser Waste (Phosphogypsum)

Yamuna Singh

Abstract Manufacturing of phosphate-based fertiliser mainly requires phosphoric acid, which is produced from phosphate ores through digesting the same by using sulphuric acid. The main waste produced in manufacturing of phosphoric acid is phosphogypsum ($CaSO_4.2H_2O$) that is moved to the identified disposal area from the plant. Based on the adopted methods of phosphoric acid production, the resultant phosphogypsum waste mainly comprises gypsum (calcium sulphate dihydrate) or calcium sulphate hemi-hydrate, besides small amounts of silica, fluoride compounds and unreacted phosphate rock. During manufacturing of phosphoric acid, REE preferentially concentrates in phosphogypsum waste. Approximately 70–85% of REE contained in phosphate rocks are transferred to phosphogypsum waste, whereas the remaining rare-earths go with the phosphoric acid. Notably, compared to REE content in ore minerals, the average content of REE in phosphogypsum waste is rather low. Nevertheless, it is significant in view of its huge quantity, as manufacturing of 1 tonne of phosphoric acid generates approximately 4.5–5.5 tonnes of phosphogypsum waste. It is possible to easily take out nearly 50% rare earths from phosphogypsum waste through leaching by sulphuric acid (0.1–0.5 M). From the leached liquor rare earths may be separated by precipitation or solvent extraction. Considerable scope is evident for value addition during complete cycle of phosphate ore processing right from its mining to application in fertilisers to agriculture to creation of wealth from its waste (phosphogypsum) and finally reusing waste as land-filling material for reclamation of land and thereby supporting afforestation, gardening and land beautification. There is a need to carry out comprehensive characterisation of phosphatic rocks and phosphorite deposits of India not only to assess their suitability for fertiliser industry, but also for the possible recovery of valuable metals from them and their waste products like phosphogypsum.

Keywords REE recovery · Valuable metals · Phosphatic rock · Phosphate fertiliser waste · Phosphogypsum

Y. Singh (✉)
Centre for Earth, Ocean and Atmospheric Sciences, University of Hyderabad, Dr. C.R. Rao Road, PO Central University, Gachibowli, Hyderabad 500 046, India
e-mail: yamunasingh2002@yahoo.co.uk

K. Randive et al. (eds.), *Innovations in Sustainable Mining*, Earth and Environmental Sciences Library, https://doi.org/10.1007/978-3-030-73796-2_9

1 Introduction

The ongoing build up in growth of demand for Rare Earth Elements (REE) is on account of sustained efforts to augment clean energy manifolds and also requirements in high-tech domains. Progressively growing demand for rare earth magnets is being driven largely by electric vehicles, wind generators, medical devices, smart phones and aerospace and defence applications. Military industrial complex depends on availability of REE due to usage of alloys with rare earth additives in combat aircraft fuselages. After imposition of supply restrictions by China research activities have been intensified all over the world to assess non-conventional REE resources, namely, industrial process residues. Notably among them include phosphogypsum, coal fly ash, incinerator ash, bauxite residue (red mud), mine tailings, metallurgical slags and waste water streams.

To mitigate the fear of non-availability of REE from the world market, and to keep sustained supply and attendant growth, countries which do not possess adequate or lack in natural rare earth and critical metal resources in their territorial domain are forced to look for secondary critical metal-bearing resources, to secure their supply. It motivates such countries for investments in technospheric mining [1–3]. Therefore, there is a need to promote technospheric mining [1, 4, 5], which, with regard to critical metals, involves at least five important steps [2, 3, 6]. Out of 5 options, end-of-life items are preferred for recovering REE by recycling [2, 7, 8]. As against this, REE-bearing industrial process residues have not been preferred much for the recovery of rare earths [2, 3] despite their availability in huge quantity. The total amounts of REE contained in such residues are adequate enough to serve as a supportive source to countries, with meagre REE resource, against their dependence on import and market price fluctuations [3]. In this chapter recovery of valuable metals (rare-earths) from phosphate fertiliser waste (e.g., phosphogypsum) generated during production of phosphoric acid and fertiliser is described.

2 Global Scenario of Rock Phosphate Resources and Mining

2.1 World Scenario

Natural resources of the rock phosphate are linked with sedimentary and igneous geological processes [9]. A major share of the global reserve is contributed by sedimentary phosphate rock (~85–90%), which occurs in Florida, Morocco and the Middle East. Whereas, remaining part of the global reserve is accounted for by igneous phosphate rock (~10–15%), and distributed mainly in the Kola Peninsula (Russia) and Brazil. The predominant phosphate mineral is apatite in bulk of the phosphate ore deposits [10]. The phosphatic rocks have received attention as a source of REE from 1960s [11]. Phosphatic rocks mostly contain LREE, especially

La, Ce, Nd, which account up to 80% of the total REE content [12]. Gambogi [13] opined that 3500 tonnes of REE would have been recovered consequent to mining and processing of 6 million tonnes (Mt) of phosphate rock in the USA during 1964. Approximately 224 Mt of rock phosphate appears to have been mined globally in 2013 [14]. Recovery of REE in the form of by-product or co-product from mining of sedimentary phosphates have also been demonstrated in several experiments [15–17]. It is surmised that about 1.4 Mt REE is contained in Western Phosphate Field, USA [18]. Significantly, it is postulated that level of REE concentration in sedimentary phosphates may help to counter global REE deficit [17]. Results of REE recovery from phosphatic rocks using 0.5 N HCl and 0.4 N H_2SO_4 as leachant are very encouraging [17]. Anticipated 2014 global individual REE (IREE) demand [19], and data of annual IREE mined from Miocene phosphate, Phosphoria Formation and Love Hollow Formation, Arkansans [17] are also available.

2.2 *Indian Scenario*

Although phosphatic deposits are known in various parts of India, most of them have not been characterised for their potentiality for possible recovery of total REE (TREE) as by-products and co-products from them. Some REE data on variants of radioactive phosphatic and associated rocks of the Bijawar basin are available [20]. The contents of rare earth in increasing abundances [21] are noted in phosphatic breccia (0.37% TREO), phosphatic sandstone (0.43% TREO), silico-phosphatic rock (0.47% TREO), phosphatic shale (0.49% TREO) and bituminous shale (0.54% TREO). Appreciable amount of rare earth is revealed by uraninite (0.49% TREO) hosted in phosphatic unit at Tummalapalle in Cuddapah basin [21].

In South Purulia Shear Zone (West Bengal), apatite mineralisation is traced over a strike length of about 100 km and a width of 4–5 km [22–24]. The area revealed approximately 12 Mt phosphatic ore, containing 11% P_2O_5 [22]. About 12,750 tonnes apatite was mined during 1992-94 [25]. With depth, minor increase in REE content is recorded [22, 38].

Extensive deposits of phosphatic rocks are present in Himalayas [26]. They carry notable concentrations of U and low rare earths (TREE 61–434 ppm), latter being dominated by La and Y. Collophane-apatite and carbonaceous matter seem to host U and REE [27, 28]. Yttrium shows enrichment in the range of 3–15 times with respect to its crustal abundance of 33 ppm Y. In Lower Tal phosphorites, TREE content is widely variable (94–789 ppm), with LREE/HREE ratios of 3–12.5 [29]. The values of TREE are not only less than the averages known for geologically old phosphorites [30], but also below world average of 700 ppm TREE [31]. Significantly, Tal phosphorites reveal enrichment of Y (19–618 ppm) compared to average shale (27 ppm in PASS), with its average content being in the range of global average of 40–6110 ppm Y for phosphorites [31]. Relative to lanthanides, preferential uptake of Y may be the reason for its higher content [29]. The phosphatic rocks from Sikkim reveal U concentration [32, 33].

3 Generation of Phosphogypsum Waste in Phosphoric Acid Production

Phosphorites and phosphatic rocks are widely used for manufacturing of phosphate-based fertilisers using a wet-process phosphoric acid (WPA), in which sulphuric acid reacts with phosphates and produces phosphoric acid and phosphogypsum [34]. Concentration of REE in phosphatic rock formed by igneous process is higher (1–2 wt%) relative to those formed by sedimentary process (0.01–0.1 wt%). For manufacturing fertilisers, indigenous sources of phosphatic rocks are inadequate in India. Accordingly, phosphatic rocks are imported by India from several countries, namely, Jordon, Morocco, Senegal, Togo, Nauru, Egypt, Algeria, Syria, Tunisia, Israel, Vietnam, China, Florida, African Phalaborwa, Brazilian and Russia. The REE abundances (0.22–3.95% RE_2O_3) in untreated phosphatic rocks from various countries reveal a large variation [36, 38]. In northern China, a few recently discovered phosphate deposits are reported to contain up to 6.41% RE_2O_3 [35]. Phosphorites and phosphatic rocks of India also show considerable variation in REE contents [36, 38].

In Indian scenario, an upward trend is noted in mining of phosphatic rocks (data in 000 tons) commencing from 806.5 during 2000-01 to 1587 during 2009-10 [36]. The main producers of phosphogypsum are given by [37, 38]. It is apparent that nearly 83200–10922885 metric tonnes of phosphogypsum was generated during the period 2006–2011 by individual industries [36]. As several studies have shown interesting outcome, phosphogypsum generated by phosphoric acid plants also appear equally attractive for recovering rare earths [38].

When phosphoric acid is made from phosphatic rock through sulphuric acid route, it results in phosphogypsum as the main by-product [39]. Phosphogypsum mainly comprises gypsum (calcium sulphate dihydrate) or calcium sulphate hemi-hydrate, besides small amounts of silica, fluoride compounds and unreacted phosphate rock. Significantly, as phosphoric acid exceeds mass, production of per tonne of phosphorous pentaoxide results in phosphogypsum to the tune of 4.5–5.5 tonnes [40], which is moved to disposal place from phosphoric acid plant [40].

4 Recovery of REE from Phosphogypsum

Out of total REE contained in phosphatic rock, nearly 70–85% REE is transferred to phosphogypsum and remaining amount accumulates in phosphoric acid [12, 35]. Phosphogypsum is reported to contain on an average 0.4 wt% REE [12]. Clearly, values are much less when compared with abundance in any REE ore minerals. Despite this, it cannot be ignored in view of enormous global mining of phosphate rock, which was nearly 224 Mt in 2013 [14]. Accordingly, phosphogypsum assumes significance for REE recovery [38].

In a leaching experiment, involving 0.1–0.5 M H_2SO_4 solution, it has been shown that approximately 50% rare earths contained in phosphogypsum may be taken into solution [12]. It is also noted that recovery of rare earth can be improved further if H_2SO_4 solution is made to percolate freely on its own through phosphogypsum stacked column [41], or agitating phosphogypsum through a mechanical device [42]. Afterwards, leached liquor may be subjected to precipitation or solvent extraction to obtain REE [5, 12]. Steps involved in recovering REE from phosphogypsum is also available [43]. A typical flowsheet is also shown (Fig. 1).

Recently, realising the potentiality of phosphogypsum as a potential source of REE, Rainbow Earths entered into an agreement with Bosveld Phosphates for the development of the Phalaborwa rare earths project jointly in South Africa [44]. The

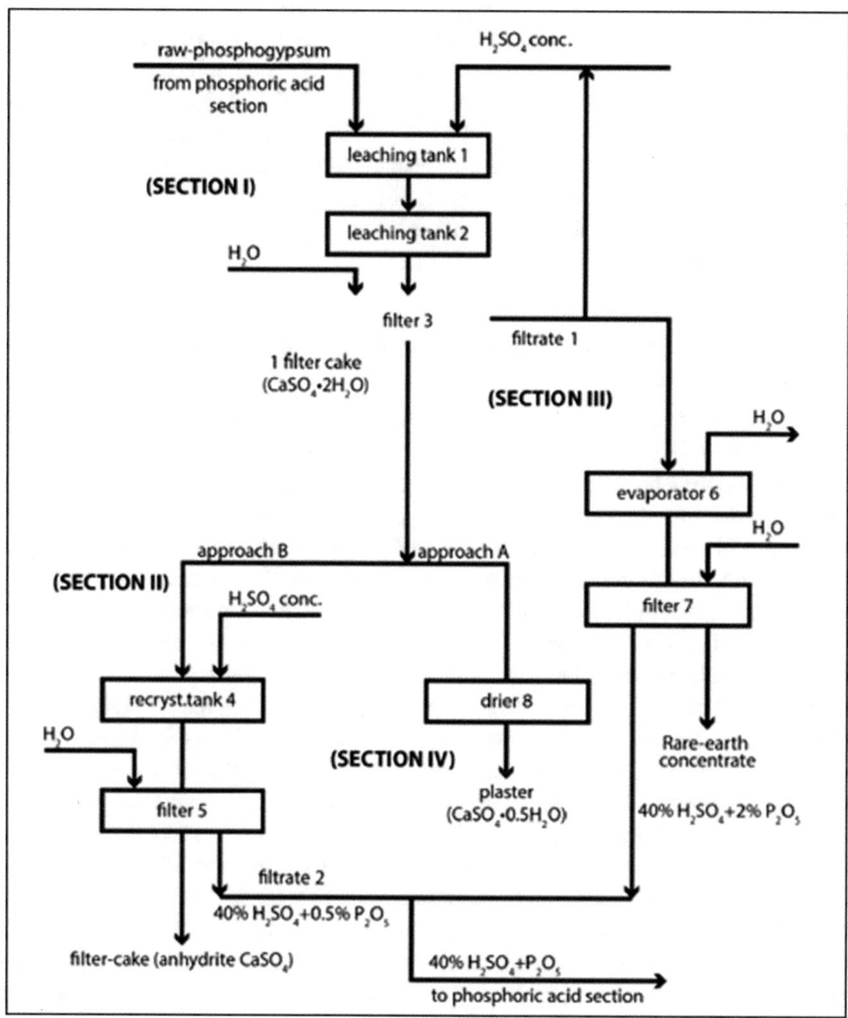

Fig. 1 Process for recovery of rare earths from phosphogypsum with production of purified anhydrite for use as plaster [5]

phosphate rock mining has resulted in nearly 35 million tonnes of phosphogypsum, containing on an average 0.60% TREO, which, on initial assessment, has indicated about 210 000 tonnes of contained TREO, with Nd and Pr put together may constitute 30% of the total REO basket [44]. Pilot plant scale studies have revealed about 80% recovery of rare earth carbonate from phosphogypsum [44].

5 Recovery of REE from Phosphoric Acid

Apart from phosphogypsum, some amounts of rare earths also go with phosphoric acid [34, 45]. Scope for recovery of rare earths from wet phosphoric acid by leaching, liquid-liquid extraction and liquid-solid extraction has been demonstrated in laboratory experiments [45, 46]. Relatively higher values of REE have been recovered when concentration of phosphoric acid is lower. It is more so with the use of kerosene diluted PC88A extractant. In this regard, various workers have developed flow sheets [35, 46, 47]. Not only as a by-product [35, 48], but also simultaneous recovery and extraction of U and REE from phosphoric acid is feasible [49].

6 Value Addition and Achievement of Circular Economy

In a study it has been estimated that with phosphate rock every year about 100,000 tons of REEs are co-mined, which are either rejected as waste or go along with phosphate-bearing fertiliser or end product [50]. Accordingly, there is enough scope for value addition in complete cycle of phosphate ore right from after mining to application in fertilisers to agriculture to creation of wealth from its waste (phosphogypsum) and finally reusing waste as land-filling material for reclamation of land and thereby supporting afforestation, gardening and land beautification. Phosphatic ores contain significant amounts of REE and U, which can be separated from raw ore, purified and refined for further uses. Flowsheet for the recovery of REE from phosphate rock, using H_2SO_4, by precipitation as hydroxides is given by Habashi [12]. Afterwards, phosphatic ore can be used for manufacturing of phosphoric acid. Phosphogypsum thus produced may be routinely subjected to REE recovery as described above. Even phosphoric acid can be processed to recover U and REE for which also flow sheets are available. It will lead to value addition in second stage. After recovery of REE from phosphogypsum, it can be used as land-filling, which will support good gardening leading to beautification and reclamation of degraded land. In addition, REEs can also be recovered from phosphate mining waste products and acid mine drainage [51–53]. In Indian context, value addition can be done not only in imported phosphorites, but Indian phosphorites also form good candidates for value addition at various stages leading to achievement of circular economy. Apart from this, phosphatic dolostone-hosted low grade uranium ore (which also contains up to 15 wt% P_2O_5) of Tummalapally – MC Pally in Cuddapah basin [54], which is under mining by M/s Uranium Corporation of India Limited, Department of Atomic Energy, Government of India may be upgraded for phosphate ore that may be used as natural substitute for phosphate-fertiliser in agriculture. Phosphate may also be recovered from beneficiation slimes [52, 55]. Also, from the waste Mo, Cu, Pb can be recovered, which are associated with carbonate ore in low concentrations, e.g., Mo 200–1260 ppm, Cu 25–670 ppm, Pb 60–290 ppm and Sr 190–570 ppm [54]. Carbonate-rich mine waste can be effectively used as de-fluoridation media for treating fluoride-rich drinking water in the adjoining regions where fluorosis is common [56].

7 Suggestions for Future Research

India is endowed with several phosphorite deposits. These have not yet been characterised for their potentiality especially for strategic and critical metals. Even REE and associated metals data is lacking on reported uraniferous carbonate rocks from the upper part of Krol Formation in Nainital Syncline in Lesser Himalaya [57]. Therefore, there is a need to carry out comprehensive characterisation of phosphatic rocks and phosphorite deposits not only for their suitability for fertiliser industry, but also for the recovery of valuable metals from them and their waste products like phosphogypsum. This work may prove rewarding in the long run in planning for possible recovery of several metals including REE from them and their waste products like phosphogypsum.

8 Summary

Manufacturing of phosphate-based fertilisers need phosphoric acid, which is manufactured from powdered phosphate ores by treating the same by sulphuric acid. Approximately 70–85% of REE contained in phosphate rocks are transferred to phosphogypsum waste, whereas the remaining rare-earths go with phosphoric acid. Although the average content of REE in phosphogypsum waste is rather low, compared to REE content in REE ore minerals, it is attractive due to huge quantity. Accordingly, resultant phosphogypsum waste forms an economic source for REE recovery. About 50% REE contained in phosphogypsum waste may be recovered by leaching using 0.1-0.5 M H_2SO_4 solution, with scope for improvement in REE recovery. The rare-earth may be taken out from the leached liquor through precipitation or solvent extraction. Considerable scope is evident for value addition during complete cycle of phosphate ore processing right after its mining to application in fertilisers to agriculture to creation of wealth from its waste (phosphogypsum) and finally reusing waste as land-filling material for reclamation of land and thereby supporting afforestation, gardening and land beautification.

Acknowledgements I am grateful to Dr. Kirtikumar Randive for inviting me to contribute this Chapter for the edited book entitled, "Innovations in Sustainable Mining"; to Dr. Dinesh Gupta and Shri Pramod Kumar for their critical review comments and valuable suggestions; and to Dr. Alexis and Dr. Femina Joshi Arul from M/s Springer for editorial comments and handling manuscript in a professional way.

References

1. Johansson N, Krook J, Eklund M, Berglund B (2013) An integrated review of concepts and initiatives for mining the technosphere: towards a new taxonomy. J Cleaner Prod 55:35–44
2. Binnemans K, Jones PT (2014) Perspectives for the recovery of rare earths from end-of-life fluorescent lamps. J Rare Earths 32:195–200
3. Binnemans K, Jones PT (2015) Rare earths and the balance problem. J Sustain Metall 1:29–38
4. Binnemans K, Pontikes Y, Jones P T, Gerven T V, Blanpain B (2013) Recovery of rare earths from industrial waste residues: a concise review. In: Proceedings of the third international slag valorisation symposium (19–20 March 2013, Leuven, Belgium), pp 191–201
5. Binnemans K, Jones PT, Blanpain B, Gerven TV, Pontikes Y (2015) Towards zero-waste valorisation of rare-earth-containing industrial process residues: a critical review. J Cleaner Prod 99:17–38
6. Jones PT, Gerven VT, Van Acker K, Geysen D, Binnemanas K, Fransaer J, Blanpain B, Mishra B, Apelian D (2011) CR3: cornerstone to the sustainable inorganic materials management (SIN2) research program at KU Leuven. JOM 63:14–15
7. Anderson CD, Anderson CG, Taylor PR (2013) Survey of recycled rare earths metallurgical processing. Can Metall Q 52:249–256
8. Innocenzi V, De Michelis I, Kopacek B, Veglio F (2014) Yttrium recovery from primary and secondary sources. A review of main hydrometallurgical processes. Waste Manag 34:1237–1250
9. Habashi F (1998) Solvent extraction in the phosphate fertiliser industry. In: EPD congress, pp 201–218
10. Rutherford PM, Dudas MJ, Samek RA (1994) Environmental impacts of phosphogypsum. Sci Total Environ 149:1–38
11. Anon. (1966) Rhodia, Umicore to recycle rare earths. Chem Eng News 89:16
12. Habashi F (1985) The recovery of the lanthanides from phosphate rock. J Chem Technol Biotechnol A 35:5–14
13. Gambogi J (2014) Mineral commodity summaries: rare earths 2014. US Geological Survey
14. Jasinski SM (2014) mineral commodity summaries: phosphate 2014. US Geological Survey
15. Christmann P (2014) A forward look into rare earth supply and demand: a role for sedimentary phosphate deposits? Proced Eng 83:19–26
16. Simandl G (2014) Geology and market-dependent significance of rare earth element resources. Miner Dep 49:889–904
17. Emsbo P, McLaughlin PI, Breit GN, du Bray EA, Koenig AE (2015) Rare earth elements in sedimentary phosphate deposits: solution to the global REE crisis? Gondwana Res 27:776–785
18. Emsbo P, McLaughlin PI, du Bray EA, Anderson ED, Vadenbroucke T, Zielinski RA (2016) Rare earth elements in sedimentary phosphorite deposits: a global assessment. Rev Econ Geol 18:101–113
19. BGS (British Geological Survey), Natural Environment Research Council (2011) Rare earth elements−mineral profile. definitions, mineralogy and deposits. www.MineralsUK.com
20. Roy M, Rawat TPS, Bhairam CL, Parihar PS (2014) Diverse mode of phosphatic rocks in the environs of Proterozoic Bijawar and Sonrai basins and its relevance to uranium mineralisation−a case study from central India. J Geol Soc India 83(3):259–272

21. Roy M, Roy M (2008) Temperature dependent textural and chemical variations in uraninites from diverse geological environments of India. J Geol Soc India 72:155–167
22. Basu SK (1993) Alkaline-carbonatite complex in the Precambrian of south Purulia shear zone, eastern India: its characteristics and mineral potentialities. Indian Min 47:179–194
23. Basu SK, Ghosh RN (1999) Rare earth elements and rare metal metallogeny in Precambrian terrains of Purulia district, West Bengal, and its status of exploration. In: Proceedings of national seminar on strategic rare and rare earth minerals and metals. Geological Survey India, Hyderabad, pp 129–140
24. Basu SK, Bhattacharyya T (2014) Petrography and mineral chemistry of alkaline-carbonatite complex in Singhbhum crustal province, Purulia region, eastern India. J Geol Soc India 83(1):54–70
25. IBM (Indian Bureau of Mines) (1996) Indian Minerals Year Book 1995, I. Indian Bureau of Mines Press, Nagpur, India, General reviews
26. Khan KF, Dar SA, Khan SA (2011) Petrography of phosphorite deposits of Durmala and Maldeota, Uttarakhand. J Indian Assoc Sedimentol 30(1):33–41
27. Saraswat AC, Balakrishnan SP, Varadaraju HN, Taneja PC, Bargaje VB (1970) Geochemical data on the uraniferous phosphorite of Mussoorie, Dehradun district, Uttar Pradesh, India. UNESCO-ECAFE Seminar on Geochemical prospecting methods and techniques, Paradeniya Ceylone, September 1970
28. Sankaran RN, Singh Y, Satyanarayana K, Shivkumar K, Mohanty BK, Krishnamurthy V (1995) Secondary sources of rare earths in India. Proc. Seminar on Recent Developments in the Science and Technology of Rare Earths, organised by IREL, Udyogamandal, Kerala, during December 14–16, pp 16–20
29. Mazumdar A, Banerjee DM, Schidlowski M, Balaram V (1999) Rare-earth elements and stable isotope geochemistry of early Cambrian chert-phosphorite assemblages from the Lower Tal Formation of the Krol Belt (Lesser Himalaya, India). Chem Geol 156:275–297
30. Kolodny Y (1981) Phosphorites. In: Emiliani C (ed) The sea, vol 7. Wiley, New York, pp 981–1023
31. Altschuler ZS (1980) The geochemistry of trace elements in marine phosphorites Part I. Characteristic abundances and enrichment. In: Bentor YK (ed) Marine phosphorites-geochemistry, occurrence, genesis. society of economic paleontologists and mineralogists, Tulsa, OK, USA, pp 19–30
32. Sagar S, Parthasarathy TN (1989) Uranium mineralisation in the Buxa carbonate-orthoquartzite sequence of South Sikkim, India. Explor Res Atomic Miner 2:39–247
33. Parthasarathy TN, Parihar PS, Krishna KVG (1990) Potash feldspar-rich alkaline rocks of Rangat tectonic window, South Sikkim: their genesis and relationship to uranium, thorium and REE mineralisation in the area. Explor Res Atomic Min 3:115–130
34. Liang H, Zhang P, Jin Z, DePaoli DW (2017) Rare-earth leaching from Florida phosphate rock in wet-process phosphoric acid production. Min Metall Proc 34(3):1–8
35. Singh DK, Kain V (2020) Phosphatic resources: a valuable wealth of rare earths. In: Critical and rare earth elements—recovery from secondary resources. CRC Press, Chapter 17, pp 331–342
36. CPCB (2014) Guidelines for management and handling of phosphogypsum generated from phosphoric acid plants. Hazardous Waste Management Series: HAZWAMS/ 2014–2015. Ministry of Environment & Forests, Parivesh Bhavan, East Arjun Nagar, Delhi—110032, June 25, pp 1–47. www.cpcb.nic.in
37. IBM (Indian Bureau of Mines) (2015) Indian Minerals Yearbook 2014 (Part-III: Mineral Reviews), 53rd (edn), Gypsum, Ministry of Mines, Government of India, Nagpur, pp 1–26
38. Singh Y (2020) Rare earth element resources: Indian context. SESS-Springer, pp 1–395
39. Koopman C, Witkamp GJ (2000) Extraction of lanthanides from the phosphoric acid production process to gain a purified gypsum and a valuable lanthanide by-product. Hydrometallurgy 58:51–60
40. El-Didamony H, Ali MM, Awwad NS, Fawzy MM, Attallah MF (2012) Treatment of phosphogypsum waste using suitable organic extractants. J Radio Anal Nucl Chem 291:907–914

41. Lokshin EP, Tareeva OA, Elizarova IP (2011) Processing of phosphodihydrate to separate rare-earth elements and obtain gypsum free from phosphates and fluorides. Russ J Appl Chem 84:1461–1469
42. Todorovsky D, Terziev A, Milanova M (1997) Influence of mechanoactivation on rare earths leaching from phosphogypsum. Hydrometallurgy 45:13–19
43. Jarosinski A, Kowalczyk J, Mazanek C (1993) Development of the Polish wasteless technology of apatite phosphogypsum utilisation with recovery of rare earths. J Alloys Compd 200:147–150
44. Bulbulia T (2020) Rainbow, Bosveld to co-develop South African rare earth project. In: Nair VR (ed), Rare Earth Assoc India (REAI) News 20(11):1–102
45. Lokshin EP, Tareeva OA, Elizarova IP (2013) Recovery of rare earth elements from wet process extraction phosphoric acid. Russian J Appl Chem 86(55):623–628
46. Reddy BR, Kumar JR (2016) Rare earths extraction, separation, and recovery from phosphoric acid media. Solvent Extr Ion Exchange 34(3):226–240
47. Radhika S, Kumar BN, Kantam ML, Reddy BR (2011) Solvent extraction and separation of rare-earths from phosphoric acid solutions with TOPS 99. Hydrometallurgy 110:50–55
48. Singh DK, Mondal S, Chakravartty JK (2016) Recovery of uranium from phosphoric acid: a review. Solvent Extr Ion Exchange 34(3):201–225
49. Bunus F, Miu I, Dumitrescu R (1994) Simultaneous recovery and separation of uranium and rare-earths from phosphoric acid in a one-cycle extraction-stripping process. Hydrometallurgy 35(3):375–389
50. Zhang P (2014) Comprehensive recovery and sustainable development of phosphate resources. Proced Eng 83:37–51
51. Felipe E, Silva G, Vidigal B, Ladeira AC (2017) Recovery of rare earth elements from acid mine drainage. In: Sustainable Industrial Processing Summit 2017, vol 1: Barrios Intl. Sym/Non-ferrous Smelting & Hydro/Electrochemical Processing; Flogen Star Outreach: Mont-Royal, QC, Canada, pp 61–64. ISBN 978-1-987820
52. Liang H, Zhang P, Jin Z, DePaoli DW (2018) Rare-earth and phosphorus leaching from a flotation tailings of Florida phosphate rock. Minerals 8:416. https://doi.org/10.3390/min8090416
53. Laurino JP, Mustacato J, Huba ZJ (2019) Rare earth element recovery from acidic extracts of Florida phosphate mining materials using chelating polymer 1-Octadecene, Polymer with 2,5-Furandione, Sodium salt. Minerals 9:477. https://doi.org/10.3390/min9080477
54. Vasudeva Rao M, Najabhushana JC, Jeyagopal AV (1989) Uranium mineralisation in the Middle Proterozoic carbonate rock of the Cuddapah Supergroup, southern Peninsular India. Explor Res Atomic Min 2:29–38
55. Zhang P, Bogan M (1995) Recovery of phosphate from Florida beneficiation slimes. Miner Eng 8:523–534
56. Cyriac B, Balaji BK, Satyanarayana K, Rai AK (2011) Studies on the use of powdered rocks and minerals for defluoridation of natural water. J Appl Geochem 13(1):62–69
57. Sinha KK, Ghosh D, Rai SD, Khandelwal MK (2016) Uranium mineralisation in carbonate rocks from the upper part of Krol Formation in Nainital syncline and its significance for uranium exploration in Lesser Himalaya. J Geol Soc India 88(3):295–304

Harnessing the Potential of Microbes for Rejuvenating Soils from Mining Sites: An Initiative for Environmental Balance and Value Addition

Anand Barapatre, Nishant Burnase, Reena Das, and Shraddha Jaiswal

Abstract Soil is a natural resource over which whole biomass thrives. It is an essential element of the human civilization since soils ensure food safety. It takes hundreds of thousand years for the natural agencies to create soils useful for cultivation. However, the anthropogenic activities such as mining have resulted in abusing such an important natural resource. The contaminated soils not only damage the green cover, but also provides pathways for movement of toxic elements generated at the mining sites. The quality of soil structure, organic matter, and micronutrients are significantly affected in the soils of active mines (often preserved in mining dumps), abandoned mines, and proximal areas. Microbial communities of soil are the key aspect in reclamation, as they play a main part in nutrient cycling, revegetation, geochemical revolutions, and soil development. The microbial diversity in mine sites consists of several species, mainly fall under the Proteobacteria, Acidobacteria, Firmicutes, and Bacteroidetes. The microbial symbionts like bacteria colonize in root and ectomycorrhizal fungi associated with local vegetation in the mine sites are extremophiles and shows very high activity. This chapter deals with the significance of different organisms which enhance soil fitness and the restoration of abandoned mines as well as in this chapter it also deals that how microbial activity and diversity can be deployed for soil rehabilitation. Other highlights include plant−microbiome interactions in plant establishment at the mine-sites and decoding the molecular mechanism of adaptation and resistance of rhizosphere and non-rhizosphere microbes at the mine site while they are involving in remediation process. At the end this chapter deals with the role of advance molecular techniques and genetically engineered organism in remediation of mining sites. Use of microbes for restoring the soil fertility provides a great value addition to the existing natural resource.

A. Barapatre (✉)
Central Instrumentation Facility, Faculty of Science, Indira Gandhi National Tribal University, Amarkantak 484887, Madhya Pradesh, India

N. Burnase · S. Jaiswal (✉)
National Institute of Miners' Health, Nagpur 440023, Maharashtra, India

R. Das
Dr. APJ Abdul Kalam Centre of Excellence in Innovation and Entrepreneurship, Dr. MGR Education and Research Institute, Chennai 600095, Tamil Nadu, India

K. Randive et al. (eds.), *Innovations in Sustainable Mining*, Earth and Environmental Sciences Library, https://doi.org/10.1007/978-3-030-73796-2_10

Keywords Phytoremediation · Mining · Rhizosphere microbes · Soil · Remediation

1 Introduction

Mining is an ancient occupation of human civilization which has been supporting greatly to the progress of mankind, however extraction of ores, minerals and other deposits are associated with the metal toxicity induced health problems all over the globe [1]. Increasing population and rapid growth in industrialization have resulted into development of mining sector which is major cause of environmental pollution all over the globe [2]. Each country has vacated mining sites accompanied with large number of wastes which contaminates soil and proximal water sources. It also produces acid mine drainage (AMD), giving rise to problems such as water acidification, phytotoxicity and damage to vegetation. Contamination arising from ongoing mining activities and abandoned mining sites is a global issue and one of the major causes of environmental contamination affecting the communities living in the surroundings [3, 4]. Mining also creates geomorphologic problems which includes landscape modification, erosion, flooding, and landslides. Health issues induced due to the inhalation of metal, mining dust and fumes have always been the greatest occupational health concern. There are various approaches of mining for extracting deposits from earth crust which can be broadly classified as underground mining and surface mining. For surface mining large part of the vegetation has to be removed to create open pits followed by blasting of rock and soil which produces tones of dust and pollute surrounding environment [5]. The metals such as cadmium (Cd), chromium (Cr), copper (Cu), lead (Pb), manganese (Mn), mercury (Hg), metalloids like arsenic (As) and minerals such as asbestos are considered as prominent pollutants of environment, exposure of these metals causes abnormalities both in animals and plants [6]. Mining activities alters the chemical, physical and biological properties of the soil, surrounding water sources and environment making it unsuitable for the growth of vegetation [7]. The microbial communities found in the soil plays a vital role in the maintenance of its structure, fertility, physical and biochemical characteristics and it varies as per the chemical composition of site [8, 9]. Bioremediation is the process of using living organism mainly includes green plants and microbes to treat and control contamination in water, soil and air [8]. Application of any particular plant or microbial species in the process of remediation is mainly depends on factors like environmental conditions, water properties, nutrient sustainability, characteristics of contaminants and soil of the respective site [10]. Microbial remediation is a common term which is used when microbes are applied for the treatment of contaminated site while the use green plants for the same purpose is known as phytoremediation. Both plants and microbes can be used to treat mining sites contaminated with metals and metalloids, some xenobiotic contaminants. The efficiency of the remediation is depending on the bioavailability of the contaminants, it includes uptake and transfer of contaminants to the microbial cells and its metabolism. The use of both synthetic and natural chelators enhances the rate of phytoremediation, it increases the extraction of several metal pollutants including Cd, Cu, Ni, Pb, U, and Zn. Phytoremediation is further classified as phytoextraction; used for metal-contaminated sites, phytostabilization; to decrease the bioavailability and mobility of metals in the environment, phytovolatilization; the conversion of contaminant in the less toxic gaseous forms, phytodegradation; uptake of complex contaminant from the rhizospheric area and its conversion into less toxic form, phytofiltration; process of removal of contaminants from the water surface, rhizoremediation; the synergistic association of plans and microbes to enhance remediation [11, 12]. The use of genetic engineering techniques in the field of bioremediation has boost the process via introducing special characteristics in the native organisms. It facilitates the use of genetically engineered organism to treat specific pollutant if naturally occurring microbes lacking ability of performing it. Some of the toxic xenobiotic contaminants such as nitrated and halogenated aromatic compounds, as well as some pesticides and explosives are chemically inert and cannot be degraded by naturally occurring microorganism efficiently; these limitations can be solved by the use of genetically modified organisms [13].

2 Types of Contamination Present at Different Mining Sites

Metals and minerals have widely been used by humans in various applications both in occupational and domestic settings. Beside the benefits of these metal elements, mining of these metals has always been a great health concern. Exposure to heavy metals and their compound are associated with the development of many health issues [14]. In this section the contaminants of mining sites which are known to be prominent threat to human and environment health are explored.

2.1 Metals

2.1.1 Cadmium

Cadmium is a transition metal mostly present in combination with zinc (Zn), copper (Cu) and lead (Pb) which is at highest environmental and human worry due to its high water-solubility and high magnitude of toxicity even at low concentration [6, 15]. Cadmium enters into the environment primarily as a byproduct of mining production mainly in Zn and Pb ore fields, leaching of heavy metals into the surrounding groundwater and also via processes like AMD at abandoned mining sites [16, 17]. In human cadmium intake is mainly via inhalation of cadmium bound air born particles while plants have high affinity of cadmium uptake resulted into its accumulation in roots and shoots [15, 16]. Cd have adverse effects in plant health, it inhibits the nitrate reductase activity and hence reduces the adsorption and transport of nitrates

[18]. In humans, inhalation of cadmium causes acute respiratory distress syndromes and pulmonary edema, renal dysfunction, anemia, osteoporosis, and bone fractures [15].

2.1.2 Chromium (Cr)

Chromium (Cr) is one of the major contaminants occurs frequently in nature as chromite ($FeCr_2O_4$) and is exist in the waste rocks, slag, and solid wastes of mining processes [19]. It is used in alloys particularly in the production of stainless steel [20]. Under environmental conditions, chromium exist in its most stable states viz; Cr(III) and Cr(VI). Industrial events are the main cause of environmental pollution of chromium, the chromium concentration in the ambient air of western India is ranges from 0.016 to 0.089 μg Cr/m^3 while in drinking water it is 4–95 μg Cr/L [21]. Chromium is an essential element in animals but not in plants. Chromium is a potent carcinogen, damages DNA via generating reactive oxygen species. Prolong occupational exposure to chromium causes asthma, bronchial inflammations, lung cancer, and inflammations of the larynx and liver [20, 22]. In plants chromium(VI) affects the seed germination and resulted into the stunted root growth, it also affects the uptake of nutrients and essential metals, such as nitrates, phosphorus, Fe, Mg, and potassium [19, 23].

2.1.3 Copper (Cu)

Copper exists in its native form and also found in oxides (cuprit and tenovite), sulfides (chalcopyrite, bornite, and chalcocite), hydroxyl carbonates (malachite and azurite), and silicates [24]. It is commonly used in the building material and also for water pipes, drain pipes and it is naturally resistant to the microbial denaturation [25]. Mining and metal processing activities, such as mills, smelters and refineries of brass as well as production of alloy like bronze are the major sources of copper pollution [26, 27]. Cu is comparatively less toxic to mammals because of its low maximum contaminant levels (MCL) in drinking water. Though Cu is an essential for many organisms, yet it can affect proteins, enzymes, and nucleic acids of the cell wall, cell membrane in some microorganisms such as algae, fungi, and bacteria. Copper is naturally resistant to microbial activities and can produce several drastic effects in living beings. It affects the aquatic ecosystem more drastically by harming bacteria, algae, fungi, plants, fish, and invertebrates [27, 28].

2.1.4 Lead (Pb)

Lead is mostly found in the mineral galena (PbS) which contributes to 86.6% of lead resource while, minerals like cerussite ($PbCO_3$) and anglesite ($PbSO_4$) also contains lead. Pb has a wide variety of applications because of its high density, malleability and resistance to corrosion. The prominent source of lead pollution is mining activities and the extraction of Pb, Ag, Zn, and Cu ores. Pb released in the environment via waste effluents and aerosols in the process of recycling of batteries, burning of coal and glass manufacturing [6, 29]. Inorganic Pb is more toxic than its organic form and is reported as potent carcinogen. In humans, intake of Pb is mostly via ingestion of contaminated food and water or by inhalation of fumes. Pb is reported to induce several forms of cancer, especially in the lungs, kidney, and brain. It can inhibit the repair of DNA and increased chromosome aberrations and acts synergistically with other mutagens [30, 31]. The accumulation of lead in plants can cause several drastic effects such as low uptake of essential minerals and nutrients, and impairment of membrane permeability and inhibition of enzymatic activities and photosynthesis [32].

2.1.5 Manganese (Mn)

Manganese is analogous to iron (Fe) in respect of chemical behavior due to their coexistence in the nature. Mn mostly exists under the ocean floors as manganese nodules and also found in metamorphic, sedimentary, and igneous rocks in different zones of the earth. During steel production Mn is implemented as additive as well as it also used as alloy accompanied with metals like chromium, aluminum, and zinc. Manganese is mainly entering in the environment via mining activities, as an anti-knock agent in fossil fuels and through metal processing activities [23]. Mn is considered as one of the major heavy metal toxicants which is associated with many health problems in humans. Prolong exposure of manganese may be resulted into neurological syndromes comparable to those of idiopathic Parkinson's disease because of its ability to cross blood brain barrier [33, 34]. Mn is an essential micronutrient in plants and participates in the photosynthetic production of oxygen in chloroplast and other vital processes but it has negative effects on the plant health in acidic lands. The drastic effects like chlorosis, necrosis of leaves and leaf rolling can be induced due to overexposure of Mn to plants [23].

2.1.6 Mercury (Hg)

Mercury (Hg) is the only metal which exists in liquid form and appears as a silvery white metal. It has application in the amalgam process which is used in the extraction of gold and silver. Mercury has wide range of applications such as it is used as a slimicide in pulp and paper industry, in batteries, dental amalgam fillings, paints, cosmetic products, antifungal agents etc. Mercury is released in the environment by artisanal gold mining, coal and oil combustion, cement production, emissions from contaminated sites etc. [35]. Methylated mercury has tendency to accumulate in the food chain via bio-magnification and it is one of the prominent neurotoxins causing severe neurological disease such as congenital cerebral palsy, and mental retardation in humans [36].

2.2 Metalloids

2.2.1 Antimony (Sb)

Antimony exists in a nature as sulfides and complex Cu-Sb, Pb-Sb, and Hg-Sb sulfides and oxides and appears as bluish-white, brittle, and glossy metalloid. It is used as additive in textiles, papers, and plastics and also as pigment in paints, as a catalyst and as a mordant [37]. The main sources of Sb contamination are waste incineration plants, smelting and refining facilities and mining [38, 39]. It has been recommended as potent toxicant and carcinogenic to humans, moreover it has been reported to inhibit the root growth in grass crops like maize and wheat [40, 41].

2.2.2 Arsenic (As)

Arsenic is primarily found in sulfide deposits as arsenides, sulfides, and sulfosalts and appears as brittle, steel-grey, crystalline metalloid. It was used as pesticides, insecticides, fungicides, and soil sterilizers in both organic and inorganic forms but due to its toxicity most of its agricultural and pharmaceutical application is banned. The primary source of arsenic contamination includes the burning of coal and oil which containing traces of arsenic. As per the report of WHO smelting processes, ore refining, waste incineration, disposal of e-waste are also involved in the arsenic contamination [23, 42]. As per International Agency for Research on Cancer (IARC) arsenic has listed as group-I carcinogens for humans. Arsenic induced toxicity is occurring manly via occupational exposure inhalation of pollutants at mining areas, waste incineration, coal-fired power plants, smelting, and battery assembly facilities. Acute poisoning of arsenic produces symptoms like nausea, vomiting and diarrhea. Arsenic has adverse effects on plants, previous studies on crops shown that it leads to reduction in the biomass of wheat, maize crops after exposure of 1.9 mg As/L [40, 43].

2.3 Minerals

2.3.1 Asbestos

Asbestos is a collective term used for the naturally existed silicate minerals with fibers and this fiber are long, thin, and flexible in nature. Asbestos can be classified into two separate groups as sheet minerals and amphibole minerals. It has been widely used in many industries due to its different properties like high stretchable strength, elasticity, corrosive resistant and thermal degradation nature, and high electrical resistance [44]. The sources of asbestosis contamination are weathering and erosion of natural deposits and asbestos-bearing rocks. Anthropogenic activities like crushing, screening, milling of ore, open-pit mining operations, extraction processing, and manufacturing asbestos are the prime causes of asbestosis contamination [45, 46]. The route of intake of asbestosis in human is ingestion of contaminated food and water and inhalation of asbestosis aerosol fibers. Asbestosis is a potent carcinogen and also associated with the occupational respiratory track disorders and inflammatory abnormalities like asbestosis [47].

3 Types and Characterization of Microbial Communities Existing at Different Mined Sites

Heavy metals are ubiquitous component of environment, but owing to anthropogenic activities over the years, there has been colossal accretion of it in soil and water. Contamination caused by heavy metals has been verified for their detrimental effects on criterion such as size, composition, apart from metabolic activities of the microbial community [48]. Apart from that, there have been reports stating deliberate damage to microbes of such environments and indirectly the organisms which are in the vicinity. Metal mining and their purification procedures like smelting and tailing constitutes major source of heavy metal pollution in soil and water [49].

Mine tailing has detrimental effect on soil biodiversity considering the dispensation of heavy metals from the minerals by intensified usage of sulphuric acid. Different microbial communities have different levels of tolerating toxicity by heavy metals [50]. For instance, Alvarez et al. [51] have reported Actinobacteria can tolerate higher concentrations of heavy metals as compared to other soil microorganisms. Microbes have derived unique methods to overcome stress created by toxic pollutants and thereby making them unsusceptible to the pernicious environment which seems vulnerable for other higher microorganisms [52]. Not much literature is available on the diversity of the acquired and unacquired microbiota of mined soils. Till date, profiling of the microbial community is a result of techniques such as 16S rRNA gene sequencing, and comparative genomic analysis for studying the genes involved for administering resistance towards various toxic elements and heavy metals [53]. Top soil environment, drainage, tailing and sediments along with rhizospheric and phyllospheric soil are reported to embrace maximum microbial activity in mined sites.

3.1 Microbial Diversity According to Mines Nature

Microbial diversity of soils and water of mines can be studied using approaches like metagenomics and metatranscriptomics of conserved sequences of 5, 16, 18, 23 and 25s ribosomal subunits [54]. The study ages as long as understanding of *Acidithiobacillus ferrooxidans*, a bacterium which converts low quality ore to copper

(Cu), which was discovered long back by ancient Romans during mine tailing activities, as it would make the fluid appear blue in color. This bacterium comprehends for 25% of Cu production in the world [55].

Recently, this particular bacterium has been genetically modified to remediate metal contaminated soil such as mercury and cadmium contamination [56]. Depending upon the nature of the metal extraction process, microbial diversity of any mine is determined. They are mostly consisting of extremophiles such as acidophiles and thermophiles, enduring harsh pH and temperature. In a study conducted in heavy metal mines of Rio Tinto, Spain, by García-Moyano and Gonzalez-Toril [57] using the technique 16s rRNA sequencing, the microbial niche was found to belong to iron-oxidizing *Acidithiobacillus ferrooxidans, Leptospirillum ferrooxidans*, and *Ferrovum myxofaciens*. In another study by Brown et al. [58], FISH technique (Fluorescence In Situ Hybridization) determined that β and γ-Proteobacteria constitute the dominant microbial communities at an acid mine drainage in Pennsylvania, US. β-Proteobacteria was found to be the most abundant class in Copper mines of China [59] and North of Wales [60] and in Lead, Zinc and Iron mines of USA [61]. Most of these bacteria are biofilm producers. Apart from β-Proteobacteria, the copper mines also had a diversity of α-Proteobacteria, γ- Proteobacteria, *Nitrospira* sp., *Firmicutes* sp., *Thermoplasma* sp., *Ferroplasma* sp., *Sulfolobus* sp., *Methanothermus* sp. [59]. An extremely acidophilic, psychrotolerant strain of class β-Proteobacteria, known as *Ferrovum myxofaciens* was reported from extremely acidic waters of Cu mine in Wales [62]. According to the author, this organism catalyses oxidative dissolution of pyrites.

Wielinga et al. [63] have found a prevailing microflora of iron and sulphur oxidizing and manganese reducing bacteria in Silver Bow creek mine tailings, Montana, USA, rich in traces of As, Cd, Cu, Fe, Mn, Pb and Zn. The iron rich Iberian Pyrite belt acidic mine waters of Spain were rich with strains of *Leptospirillum ferrooxidans, Acidithiobacillus ferrooxidans, Acidiphilium* sp., *Actinobacteria, Acidimicrobiales, Ferroplasma* sp., *Acidobacteria* and *Ferrovum* sp. with high activity of iron oxidation and reduction [64]. Ni, Co, Cu, Zn, Pb, As, Cd and U rich Lusatian acidic lignite mine in Germany is reported to have an extreme environment of pH 1.0 and temperature above 40 °C, the microbial diversity is supposed to be fill with extremophiles such as *Ferrovum myxofaciens, Sideroxydans* sp., *Albidiferax ferrireducens, Pelobacter propionicus, Geobacter chapellei, Geobacter psychrophilus, Acidimicrobium ferrooxidans* and *Chlorobium ferrooxidans* [65]. Table 1 portrays the microbial diversity of different mined sites. In a metagenomic study conducted by Navarro-Nova et al. [66], a divergent profile of 12 apparent bacterial phyla were reported. All the strains had an apparent approach for uptake of nutrition and metabolism of heavy metals. The manifestations of pervasiveness of unique microbial diversity present in the mines conclude the fact that the microbial communities can very well adapt to the extreme conditions and metabolize heavy metals.

Table 1 Portrays the microbial diversity of different mined sites

S. no.	Mine sites	Heavy metals present in the sites	Microorganisms present	References
1.	Heavy metal mines, Rio Tinto, Spain	Fe, Zn, Cu, Ni, Mn	*Acidithiobacillus ferrooxidans*, *Leptospirillum ferrooxidans*, *Ferrovum myxofaciens*	[57]
2.	Acid mine drainage, Pennsylvania, US	Fe, Al, Mn, Cu	β and γ-Proteobacteria	[58]
3.	Copper mines of China	Fe, Cu, S, Ca	β-Proteobacteria	[59]
4.	North of Wales	Fe, S, Mn, Cu, Zn	β-Proteobacteria, *Nitrospira*, *Firmicutes*, *Thermoplasma* sp., *Ferroplasma* sp., *Sulfolobus* sp., *Methanothermus*	[60]
5.	Dyffryn Adda AMD, UK	Zn, Cu, Al, Fe, Mn	*Acidithiobacillus ferrivorans*, *Ferrovum myxofaciens*, *Acidithrix ferrooxidans*, *Acidocella* sp., *Acidithiobacillus ferrooxidans*, *Leptospirillum ferrooxidans*	[61]
6.	Acidic Copper mine, Wales	Fe, S, Mn, Cu, Zn	β-Proteobacteria, *Ferrovum myxofaciens*	[62]
7.	Silver Bow creek mine tailings, Montana, USA	As, Cd, Cu, Fe, Mn, Pb, Zn	Iron and sulphur oxidizing and manganese reducing bacteria	[63]
8.	Lusatian acidic lignite mine, Germany	Ni, Co, Cu, Zn, Pb, As, Cd, U	*Ferrovum myxofaciens*, *Sideroxydans* sp., *Albidiferax ferrireducens*, *Pelobacter propionicus*, *Geobacter chapellei*, *Geobacter psychrophilus*, *Acidimicrobium ferrooxidans*, *Chlorobium ferrooxidans*	[65]

3.2 Plant-Associated Microbial Communities in Different Mining Sites

The microbial communities associated with plants are grouped into rhizosphere, phyllosphere and endosphere. Rhizopsheric soil is rich in microbial diversity which is mostly inveigled by impeachment of plant mucilage and root exudates [67]. Phyllosphere is the aerial surfaces with least microbial diversity [68]. Heavy metals are mostly concentrated in roots and shoots [69].

Rhizospheric fungi and bacteria are mostly ectomycorrhizal, and their diversity is more abundant as compared to regular soil [70]. Plant associated microbial communities play a major role in creating favorable ramifications for plants, thus forming a crucial factor in biogeochemical cycle and in chemical signaling for improvement of vegetation in mines. Metalliferous plant species such as have been reported to exist in a lead mine at Peak, UK [71].

The plant growth promoting rhizobacteria (PGPR) like ectomycorrhizal (root colonizing) and arbuscular mycorrhizal fungi, cater crucial roles in cycling of heavy metals and nutrients and elements such as nitrogen, phosphorus and sulphur. According to a study reported by Khan et al. [72] mycorrhizal fungi provide protection against toxic substances and improve soil quality. Other bacteria reported to improve soil quality belongs to genus *Azotobacter*, *Clostridium* and *Frankia* [73]. In metal contaminated soils, the microbiome and heavy metals maintains equilibrium for survival of certain grass species such as *Andropogon gerardii* and *Festuca arundinaceae* [74].

Symbiotic relationship of plant microbiome in extreme environments is reported only for few species. Baker et al. [75] and Ghnava et al. [76] have noted endosymbiotic relationship between protists and bacteria in acidic mines of Iron Mountain, California for symbiotic fungus *Sinorhizobium meliloti* from a mined site, which could do adsorption of cadmium. Apart from heavy metals, biotic and abiotic factors like fly ash, high pH and hydrophobicity are found to dissuade plant growth and colonization of microbes in the rhizosphere. Nonetheless, willows (*Salix* spp.) could grow on fly ash contaminated soil with the help of site adapted rhizopsheric microbes [77].

In mine tailings of Northwestern Province of China which is rich in heavy metals such as Cu, Zn, Ni, Cr, rich microbial diversity of *Mesorhizobium amorphae, Sinorhizobium meliloti, Agrobacterium tumefaciens* was found to be in symbiotic association with plants such as *Robinia pseudoacacia* and *Medica golupulina* [52]. Similarly, in the mines of Santa Maria and San Francisco, Mexico, wherein tailings of heavy metals such as Zn, Cd, Pb Cu, Ni, Cb, takes place, Rhizobacteria and Arbuscular mycorrhizal (AM) fungal species maintains efficiency of plant species such as *Pteridium* sp., *Aster gymnocephalus, Gnaphalium* sp., *Crotalaria pumila* [69]. Furthermore, some ferns viz., *Equisetum debile Roxb., Pteris radicans, Nephrolepis hirsutula* are reported to have symbiotic associations with AM fungal species such as *Scutelospora* sp., *Glomus* sp., *Claroideoglomus* sp., *Acaulospora* sp. in the gold mines of Papua, Indonesia [78].

4 Interactions and Communication Between Plant and Microbes at Mining Sites

Plant microbe interaction signalling in rhizospheric microbiome constitutes the basis of the remediation of heavy metal contaminated sites. The signalling pathways frame up phytoremediation which utilizes the translocation mechanism for remediation, and rhizoremediation wherein microbes grow in the rhizospheric soil, which is in turn enhanced by root exudates, and help in remediation of pollutants [79]. Heavy metal bioavailability resolute mostly by the endemic microbial flora present in the rhizosphere. It influences the biophysical and chemical characteristics also of soil. The rhizospheric heterotrophic microorganisms solubilize and fix the heavy metals at the root systems, as well as release organic carbon as CO_2 to the atmosphere, thereby subsidizing the greenhouse effect [80]. Fungi are known to break down the contaminants by establishing a hyphal network. *Glomus* sp. along with *Trichoderma koningii* has proven to help the growth of *Eucalyptus globulus* in heavy metal contaminated sites [81].

5 Factors Involved and Affect the Bioremediation Process

5.1 Characteristic of Plant Species Used for Phytoremediation

Metal accumulation, translocation capacities, and phytoremediation efficiencies differ from plant-to-plant species. Both monocot and dicot show phytoremediation ability for different metals in a different fashion like copper is accumulate by both monocot and dicot in their roots, while lead is mainly accumulated by monocot in their roots [82]. Based on the extraction of heavy metals, two phytoextraction strategies are mainly applied in which first is hyperaccumulators while another is high biomass yielding plant type with the combination of metal-solubilizing agents. Hyperaccumulator plants, like *Thlaspi caerulescens* and *Alyssum bertolonii*, have ability to accumulate heavy metals in high concentration, but they produce very low biomass and are the limit for few metals. But high biomass producing plant can absorb an extensive range of heavy metals but at small concentration [83].

5.2 Characteristics of the Medium

The soil property is one of the major factors which can distress the metal movement and phytoavailabiliy, which ultimately affects the metal uptake ability of the plant. The soil pH plays a key influencing feature in the bioavailability of metals in the soil to plant uptake. Generally, as the soil pH value decreases, the amount of heavy metals in soil solution might be increased because of the increase in H^+ ion concentration. This increased acidity of soil will directly surge the transferrable capacity among heavy metal cations and H^+ ion which is attached on the soil particle surface. Thus, at lower soil pH i.e., in acidic soil, a huge number of heavy metal ions are released from the surface of colloids and clay mineral particles, and are free to go in soil solution [82, 83].

5.3 *The Rhizospheric Region*

The rhizospheric region of plant may also generate a great impact on the movement of heavy metals as it accumulates more metals than the neighbouring residues. The microorganisms found in the vicinity around rhizospheric region (mainly mycorrhizal fungi and bacteria) also potentially increase the availability of heavy metals in soil. The plant growth-promoting rhizobacteria (PGPR) plays an important part in plant vigoro and nutrition through various mechanisms, like siderophores formation and release (for sequestering the iron ion), phytohormones production, and bacteria connected to plant growth-promoting (PGP) activity, can improve several stresses. Root secreted substances may also play an important role in defining the competency of metal phytoremediation. The root secretes can supply nutrients for rhizosphere microorganisms and these microorganisms may also affect the mobilization and phytoavailability of heavy metals by enlightening the solubility of heavy metal ions [82, 84].

5.4 *Bioavailability of Heavy Metals*

The bioavailability of the metals is another main requirement for metal phytoremediation. A plant is unable to take up metal from the rhizosphere if the metal is not at least slightly soluble state. When the metal is in the soluble fraction in soil than only it can be taken up by plants. Available, unavailable, and exchangeable fractions are the three main category of the metal based on their accessibility for the plant. Free metal ions as well as soluble metal complexes present in soil solution known as bioavailable metals and these can also be taken up to inorganic soil constituents at ion-exchange sites. While unavailable metals are present in the soil as different fractions, like the portion chemically bound to organic materials, present as a form of oxides, hydroxides, and carbonates, and also implanted in the structure of the silicate minerals, and hence are very difficult to be absorbed by a plant. Exchangeable fractions are stand between the above two types, where it bound to organic matter, carbonates, or iron-manganese (Fe–Mn) oxides, which are not partly take out by the plant. Metals like zinc and cadmium present mainly in an exchangeable and readily bioavailable form, while lead occurs as a less bioavailable form due to present in the soil as precipitates [83].

5.5 Environmental Conditions

Some plants have ability to convert heavy metals from the immobilized form to the movable fraction, and thus they straight increase the gathering of heavy metals and their bioavailability as well as their mobility. For example, Fe^{3+} and $Mn^{3+/4+}$ oxides offer substantial bindings positions for heavy metals under aerobic conditions. Likewise, sulphide offers dynamic binding sites for copper and lead under anoxic conditions. In saltmarsh sediments, ecological situations affect the movement of heavy metals that are influenced by biogeochemical processes [83, 84].

5.6 Chelator Agents

Chelates are mainly applied in heavy metal contaminated soil to increase the phytoextraction of several metal pollutants including cadmium, cooper, nikal, lead, uranium and zinc. The main hurdle for phytoremediation of several metals, such as lead, chromium and urenium, is their presence in the soil as insoluble precipitates, which are immobile by which their phytoextraction rates are affected by their dissolvability in a soil solution [85]. For this reason, "chelate-induced hyperaccumulation" strategy is adopted by using some plants to accumulate heavy metals. In this strategy, at the contamination site specific biomass for example any short life agronomic crop like corn will grow and when enough biomass is produced, chelating materials are added to the field soil for promoting the sequestering of large amounts of metal. By these strategies, heavy metals are captured by the plant in high concentration and after death; plants are removed from the site. Metal chelating agents such as synthetics (ammonium thiocyanate, diethylene glycol mono-buthyl ether acetate (EDGA), ethylene diamine tetraethyl acetate (EDTA), HEDTA, trans-1, 2- Cyclohexane diamine tetra acetic acid (CDTA), diethylene triaminepenta acetic acid (DTPA), nitriletriacetic acid (NTA), [N, N]-bis glutamic acid (GLDA), [S, S]- ethylenediamine di-succinic acid (EDDS), citric acid (CA), oxalate, malic acid, succinate, tartrate, phthalate, salicylate, acetate, and hydroxylamine, etc.) and natural zeolites were commonly used to enhance the speed and quantity of metal uptake by the plant [85–87]. Table 2 represents some reported studies where they use a different type of synthetic and natural chelators for the removal of heavy metals.

6 Types of Remediation Processes Occur at Mining Sites

The history of the extraction of metal is very old and from prehistoric times humans extract lots of metal to fulfil their daily life requirements. At present, it has been calculated that more than 1,150 million tons of metals including heavy metals like copper, lead zinc, cadmium, mercury etc. were extracted from metal mines. But generally, a

Table 2 Study reported for the use of synthetic and natural chelator to enhance the phytoremediation process

S. no.	Plant	Metal remediate	Chelator used	Removal quantity	References
1.	*Helianthus annuus*	Cd, Cr and Ni	EDTA and HEDTA	Total metal uptake under influence of EDTA 59 g/plant; HEDTA 42 g/plant; Cd 115 mg/kg plant; Ni 117 mg/kg plant	[88]
2.	*Solanum nigrum*	Cd and Pb	EDTA	124.6 mg Cd/g of dry weight in leaves; 112.4 and 161.8 mg/kg in the shoots and roots	[89]
3.	*Phytolacca americana L.*	Lanthanum (La)	Aspartic acid, Asparagine	In whole-plant La content increased by 449 μg and 139 μg	[90]
4.	*Brassica juncea (L.) Czern. et Coss*	Uranium (U)	Citric acid (CA; 2.5 mmol), EDTA	In the roots 776 ppb (dry mass basis) in shoots 350 ppb under influence of CA; 759 for roots and 340 ppb for shoots respectively	[91]
5.	*Eucalyptus globulus & Cicer arietinum*	Lead (Pb)	EDTA	700 and 324 mg/kg	[85]
6.	*Vetiveria zizanoides L.*	Plutonium (Pu)	CA and DTPA	Removal level of 66.2% from test solution (Initial activity 100 Bq/mL, total activity 2000 Bq) after 25 days	[92]

(continued)

very small quantity (less than 2%), were extracted, processed, and produced to the desired valuable metal from their respective ores, while the rest about 97–98% is discharged as wastes from beginning to end product process in areas mainly nearby the mine operation [4]. Conferring to a report on production of mineral by International Organizing Committee (IOC), for the World Mining Congress, India ranked 4th

Table 2 (continued)

S. no.	Plant	Metal remediate	Chelator used	Removal quantity	References
7.	*Amaranthus hypochondriacus L.*	Cadmium	GLDA, NTA, EDDS, CA (alone and in combination); reported highest for GLDA and NTA combination	For Cd low (2.12 mg/kg) contaminated soil 11.87 g/pot in shoot and 2.63 g/pot in the root; for Cd high (2.89 mg/kg) contaminated soil 14.05 g/pot in shoot and 3.42 g/pot in root	[93]

amongst the mineral producer countries based on the volume of production. However, it ranked 8th position based on the value of mineral production during 2009 [94]. Globally, metal contaminated soils produce a great risk for human health as well as for the environment because of their non-biodegradable nature. Metal contamination in different parts of the world including India is primarily due to different occupational activities and it is estimated that a huge financial budget (about $1.7 trillion) is to be spent to remediate all contaminated sites alone in the USA. The occupational activities (including their end products) of paint, locomotive industry, battery business industries, printing circuits, plating, mining, petroleum, and chemical engineering industries, are the main sources of the release of heavy metal into soil and water. The World Health Organization (WHO), Occupational Safety and Health Administration (OSHA), and other country-specific agency (Bureau of Indian Standards) set a permissible limit at all over the globe for each metal for drinking water, food, and occupational working area; beyond that permissible limit, they cause numerous disorders and diseases. These heavy metals (arsenic, mercury, cadmium, chromium, copper, lead, zinc, nickel, and manganese) cause numerous health issues in humans such as damage of different organs (like liver, lungs, and kidney and nervous system), headache, asthma, allergy, cancer, heart attack, anxiety, depression, digestive problems, autoimmune diseases, vomiting, muscle weakness by entering into the food chain and bioaccumulate in different parts of the body such as muscle, bones, brain, and kidney [95–97].

In mining areas, mining not only affects the quality of air and water, but also it damages the biodiversity, and land quality by numerous mining activities, like blasting, drilling, and storage of overburden dump materials, land cleaning, construction of subsidiary amenities, and vehicles movement. The end result of the mining will lead to vegetation destruction, decreases the amount of soil essential nutrients and organic matter, diminish different biological actions, and the soil efficiency. Soil microorganisms play a very important role in the retrieval of such contaminated and affected lands, as well as soil fertility and plant health [4]. From last few decades, several metal removal methods like membrane separation, precipitation by chemical, reverse osmosis, extraction by different solvent, membrane filtration, flocculation, electrodialysis, electrochemical treatments, ion-exchange, advanced oxidation process, and by using nanotechnology adsorption (physical, chemical, and bio-adsorption) are used for the treatment of contaminated soils and water. The above methods are mainly categorised as in situ immobilization of pollutants, ex situ soil excavation and treatment and the last one is degradation/detoxification of organic and inorganic pollutants by physical, chemical, or biological means. But due to some limitations, and disadvantages in these techniques, many of these could not be used for the removal of contaminants from soil and water. These limitations are the complexity of the technique, high energy consumption, high costing, limited selectivity, fouling formation, high slime generation, and some techniques that are not applicable when the concentration of contaminants is high [95, 97]. In the above-mentioned methods till date, bioremediation process is the easiest, economic, and eco-friendly strategy for the removal of contaminants from soil and water. The estimated cost for the bioremediation for metal contaminants is around $80 per cubic yard, while it reaches $250 per cubic yard with other technologies. For lead contamination area, phytoremediation cost $50,000 per acre while through other technology cost goes up to 2.5 times i.e., $1,20,000 per acre. For decontamination of organic as well as inorganic contaminants, bioremediation propose the opportunity of degrading, eliminating, altering, immobilizing, or otherwise purifying various xenobiotics from the environment through microbes and plants possessing an extensive variety of enzyme mediated catabolic reactions. Because bioremediation worked on natural attenuation, the public considers it more acceptable than other technologies [4, 98].

6.1 Phytoremediation

Phytoremediation is the process where the green plants are planned to use, in association with variety of soil microbes, to remediate the lethal pollutants via degradation and detoxification process. The term "phytoremediation" originates from two words, i.e., "*Phyto*" (Greek word; meaning *plant*) and "*remedium*" (a Latin word; means to *correct or remove a malicious thing*), which applied mainly for the green restoration of the contaminated site by different recalcitrant organic pollutants, heavy metals etc. [99, 100]. Plants have a great and unique ability to take up the pollutants from the soil environment and achieve their decontamination by different mechanisms like phytoextraction, phytostabilization, phytovolatilization, rhizodegradation, phytodegradation, and rhizofiltration. Each mechanism having its unique properties and each of the mechanisms cannot be useful for the removal of all the pollutants [99]. The encouraging aspects of phytoremediation over other remediation process are as follows:

1. It is cost-effective and economic, as compared to other remediation techniques;
2. Eco-friendly technique having minimal disturbance properties in the environment;

3. It can be applied for almost all of the heavy metals, and
4. Effortless to use and accepted by the public [86].

Phytoremediation also has some limitations over the above advantages.

1. the limitation of heavy metal tolerated species which are little in number;
2. the limited specificity of most of these plants abide towards specific heavy metals;
3. production of comparatively less biomass due to the small size and minimal growth
4. geological and climatic limitations for plant variety, most of them can be used in their natural habitats alone;
5. relatively slow process in comparison to chemical and physical techniques, and it took long time to decrease metal contamination from the soil to a harmless and satisfactory level [86].

Therefore, the use of plants in phytoremediation is restricted due to plants species that are used for phytoremediation should:

1. have high ability to take up the contaminants;
2. have a high ability to cultivate fast in a polluted condition, and
3. to be maintain and regulate effortlessly [86].

6.1.1 Phytoextraction

Nowadays, phytoextraction is the main type of phytoremediation which largely used in metal-contaminated sites. In phytoextraction, the metals are taken-up from contaminated soil or water through plant roots and translocate and accumulation in aerial part of the plant i.e., shoot and leaves. This process is also recognized as phytoaccumulation, phytoabsorption, or phytosequestration. Different plant species generally have the ability to tolerate a high level of metals concentration, that's why they used. Because harvesting a root biomass is generally a non-feasible task and hence, plant used for the phytoextraction should possess an essential biochemical process that translocate metal to different part of the shoot [79, 99, 101]. As soon as plants are exposed to non-toxic levels of metals, gradually and sequentially plants start growing and accumulating the metals. Metals like copper, zinc, chromium, lead and nickel have been successfully reported for phytoextraction. Hyperaccumulator plants are primarily used for the phytoextraction of heavy metals which have an ability to collect the heavy metals in their branches and leave to levels higher than non-accumulating plants. The main family plants reported for hyperaccumulator include Asteraceae, Brassicaceae, Caryophyllaceae, Fabaceae, Flacourtaceae, Lamiaceae, Poaceae, Violaceae, and Euphorbiaceae [79, 86]. Plants used in phytoextraction should preferably have the following features [99, 102]:

1. They should have a high growing rate and generation of more shoot biomass
2. Good translocation capability of the collected metals from roots to shoots.

3. Should have an extensively scattered and highly branched root system.
4. High accumulation capability and high tolerance limit for target metals from soil.
5. Easy cultivation and harvesting property.
6. Upright adaptation capability for environmental and climatic conditions.
7. Resistance towards pathogens and pests as well as good repulsion property for herbivores to evade the food chain contamination.
8. Allow several harvests in a single growing cycle.
9. Association of symbiotic mycorrhizal fungi.

6.1.2 Phytostabilization

Phytostabilization is a technique which decreases the movement and bioavailability of metals in the environment. In phytostabilization, plant roots arrest the heavy metals in soils by the sorption, precipitation, complexation, or reduction in heavy metal valency at the rhizosphere area. In this process, the rhizobium and endophytic soil microbes performance an important part for the improvement of phytostabilization. Heavy metals interact with different side chain functional groups like amine and amide groups, sulfonate, sulfhydryl, hydroxyl, and carboxyl, which results in the hold of metal ions and limits their entrance into the root. They also bond with extracellular polymers made up of polysaccharides and protein moiety which remove the metals by chelating them [79, 103]. This process stabilizes the metals to prevent their movement into under-ground water or their entrance into the food chain thus decrease the risk to human health and the environment. Phytostabilization, not only remediate the polluted soils but also reduce the pollution of adjacent area and the deactivation of requisitioning of metal pollutants. Unlike phytoextraction or hyperaccumulation of metals into shoot/root tissues, phytostabilization principally emphases on the seizure and inactivation of the movement of the metals near the root soil and plant root but not inside the plant tissues [79, 99, 102].

Mine tailings or mill tailings, is one of the biggest problems at the mine site and are mainly generated by the materials residual after abstraction and ore processing. Throughout the world, these mine tailings dumping spots are prevalent from either inactive or abandoned mine sites. The global level effect of such mine tailings discarding spots is massive, because these mining sites are usually without vegetation for numerous decades, and these tailings material can spread nears area of air scattering and water erosion. Phytostabilization is one of the effective solutions of such type of metal contamination (mine tailings) and in arid and semiarid environments the use of drought-, salt-, and metal-tolerant plants effectively for the holding of heavy metals [60, 104]. Plant use for phytostabilization preferably should be native to that particular area in which this approach will applied, because they have already evolved and developed survival mechanisms appropriate to the severe climate of affected mine sites. And the secondary criteria for the use of in situ plants is to avoid the introduction of the foreigner and potentially offensive species so that regional plant diversity will not affect [60, 103].

6.1.3 Phytovolatilization

Phytovolatilization is the process in which plant absorbs metals in their elemental form and transform them into gaseous species within plant, and finally release them into the atmosphere in less toxic form. Apart from the other phytoremediation methods, it can be considered that phytovolatilization is the enduring site solution because the converted gaseous volatilized products are never re-deposited at or near the site and there is also no trace will be seen about the relocation of contaminants to other places. The major disadvantage of this method is that it should not be applied near high-density population cities or the places which having infrequent weather patterns because of such places might encourage quick release of volatile elements and there is a chance to come down with rainfall and then re-enter into the ecosystem [102, 103].

Some of the toxic metals like selenium, arsenic, and mercury can be transformed to volatile forms like dimethyl selenide and mercuric oxide by plants like *Chara canescens* (musk-grass) and *Brassica juncea* (Indian mustard) and further evaporated or volatilized into the atmosphere but these converted compounds are also toxic to human [103]. Phytovolatilization is divided into two types [105]:

1. Direct Phytovolatilization: In this type, contaminants/metals were uptake and translocation by the plant to the shoot portion and diffuse through tissue barriers like cutin or suberin present in dermal tissues.

2. Indirect Phytovolatilization: By applying the sufficient quantity of soil plants, which take a huge amount of water so that activities of plant roots may increase the fluidity of volatile impurities from the subsurface through the following ways:

- Dropping of the water table.
- Water table variations produce gas fluxes.
- By increasing the soil penetrability.
- Hydraulic rearrangement.
- Interference of rainfall that would otherwise penetrate to dilute and advert VOCs away from the surface.

6.1.4 Phytodegradation

Phytodegradation is a procedure where plants absorbed the pollutants from the rhizospheric region and breakdown it down to simple, small and less toxic forms. Plant breakdown these contaminants through two ways:

1. by metabolic process in the plant, and
2. by plant produced enzymes.

In phytodegradation, pollutants like solvents (chlorinated), insecticides, organic complexes and various inorganic compounds can be degraded. This degrades these contaminants into smaller and simpler products that are used by the plant for their faster growth [99, 103]. The plant does not involve in the heavy metal degradation process because metal is non-degradable.

6.1.5 Phytofilteration

Phytofiltration is a process of removal of pollutants from polluted surface waters or wastewaters by plants and limiting their movement to underground waters. Phytofiltration can be performed by the plant in a different manner/types. In rhizofiltration plants use roots or in blastofiltration, it is done by the plant seedlings, while in caulofiltration, caulis mean shoot perform this process. In phytofiltration, the contaminants are absorbed or adsorbed by the plant roots so that heavy metals movement were cease/minimised into the underground waters. This approach might be directed in situ, where floras are cultivated straight in the contaminated water body, which decreases costs [99, 102].

Blastofiltration takes advantage of other, in the way of the rapid rises in surface to volume ratio by which after germination and also many saplings are adsorbed or absorb large quantities of metal, creation them exceptionally appropriate for water remediation. Several advantages exist in blastofilteration like sapling cultures can be grown in different environmental parameters, such as in light or darkness; cultures are reasonable, needing only seeds, water, and air-land plant seeds are typically common, easy to obtain, cheap, and are thrown away as waste, which makes this technology even more sustainable and green [102]. Rhizofiltration is a technique were it mainly applied using aquatic macrophytes and some earthly plants are also capable to perform by using a root biofilter, formed by microorganisms to absorb, concentrate and precipitate the metals [102, 106].

6.2 Rhizoremediation

Rhizoremediation is one of the sophisticated forms of bioremediation that inspires plants to create a mutual relationship with indigenous soil rhizospheric microbial communities where plant support the microbial growth as well as stimulate them for pollutant degradation [97]. The term rhizoremediation was first utilized in 1998 during the study of the degradation trichloroethylene (TCE) in the wheat rhizosphere by bacteria expressing a stable, chromosomally encoded toluene ortho-monooxygenase (TOM). “Rhizo” originated from the ancient Greek word *rhíza*, which refers to *the root*, and “remediation” denotes *the method of degrading the contamination or unwanted material* [96]. In rhizoremediation, the microbes degrade/fix the xenobiotics/contaminants whereas plant roots deliver a niche to the microorganism and provide key nutrients. In the rhizospheric region, plant-derived substrates, like organic acids, certain monosaccharides, sterols, ions stimulate/enhance the microbe's degradation or transformation ability towards contaminations. This enhanced capability may be due to the larger number of gathered microorganisms or the accessibility of growth-supporting substrates for co-metabolism and this process also paybacks the plants as nitrogen nutrient because certain microbes help in nitrogen fixation in the roots while some others preventing the attack of other pathogenic organisms. The plant roots also provide a large surface area for microbes propagation, biofilm formation, and facilitate oxygen exchange [95–97].

In metal mining sites, increased metal concentration, supress the microbial growth, microbial biomass and essential microbial activities inhibiting the bacterial and fungal population which permanently alters the microbial community structure. Rhizosphereic bacteria dependent on plants which are resident in metal contaminated soils support the microbial populations which are:

1. resistance against different metals;
2. seize/bioaccumulate/biosorb metals into the cells;
3. have ability to convert the lethal metal species to non-lethal species by the use of different enzymes and secreting substance like exopolysaccharides.

The structure of rhizospheric microbial community is depend on several factors like, composition of the root secreted substrate, type of the plant species, type of root, plant age, and also the soil type [83, 95]. Form a long period, a diverse group of free-living soil bacteria known as plant-growth-promoting rhizobacteria (PGPR) was mainly used for assisting plants to improve plant growth and development by up taking the nutrients from the environment and preventing plant diseases developed by the metals. The PGPR linked with plants, grown in metal contaminated soils are *Azotobacter chroococcum* HKN-5, *Bacillus aryabhattai* (MCC 3374), *B. megaterium* HKP-1, *B. mucilaginosus* HKK-1, *B. subtilis* SJ-101, *Brevundimonas* sp. KR013, *Enterobacter ludwgii* (HG2), *E. aerogenes* (MCC 3092), *Klebsiella pneumonia* K5, *Paenibacillus* sp. ISTP10, *Pseudomonas fluorescens* CR3, *Rhizobium leguminosarum bv. trifolii* NZP561, *Kluyvera ascorbata* SUD165 are reported in bioremediation. Different hyperaccumulators plant linked with rhizobacteria like *B. subtilis, B. pumilus, Pseudomonas pseudoalcaligenes,* and *Brevibacterium halotolerans* are also widely used in rhizoremediation of multi-metal contaminated sites [107–109].

6.3 Bioventing

Bioventing is the utmost widespread in situ bioremediation system wherein native aerobic microorganisms are used to degrade the subsurface contaminants in the soil. At the point of contamination or pollution above water level, air and nutrients are supplied through wells. In aerobic condition the native microorganisms reduce the contaminants and the extraction well is safely vented after removal of degraded pollutant. It also recruits the low air supply which ensures amount of oxygen needed for bioremediation. The injected oxygenates air into the contaminated soil ensure the biodegradation of contaminants. For remediation of petroleum-contaminated soil, bioventing has been known to be highly effective [110].

6.4 *Bioaugmentation*

Specific microbial cultures, intrinsic or exogenous are supplemented to the contaminated site to improve the degradation of contaminated compounds when native soil microorganisms are incapable to degrade the complex pollutants. This process enhances the metabolic competence of indigenous microbial population as well as elevates the rate or level of degradation of the contaminants. A consortium of different potential microorganisms is introduced to the contaminated site which enhances intrinsic bioremediation process and ability. Mostly, general strains are preferred over the superior strains of genetically engineered microorganisms (GEMs). The direct use of GEMs may be the cause of several health hazards [111, 112].

6.5 *Biosparging*

In biosparging, pure oxygen and nutrients (if needed) are inoculated into applied zone to elevate the biotic activity of naturally occurring microorganisms. The efficacies of biosparging depend on two key features viz: soil porousness, which regulates bioavailability of contaminant for microorganisms, and biodegradability of contaminant [4]. Biosparging encourages biodegradation and procedure is similar to bioventing. Biosparging is most effectively applied at the spot were medium weight petroleum products such as diesel and kerosene, contamination occur.

7 Strategies to Use of Rhizosphere Microorganisms in Recovery of Mined Sites

Microbes found to be scarcely multiplied and flourish in metal-contaminated soils in field conditions [109]. Therefore, at the ground level, indigenous microorganisms associated with plants are important in bioremediation technologies [112]. In past few years as related with fungi, the remediation mechanisms at the molecular level are better understood in numerous bacterial species which are especially isolated from wastewater treatment units. The biochemical cue is responsible to degrade and detoxify natural toxins by microorganisms. [112, 113].

1. Addition of sewage waste into the top soil layer to stimulate growth of nitrogen fixing and ammonia oxidizing bacteria [114];
2. A consortium of plant growth promoting rhizobacteria (PGPR), filamentous fungi, and N_2 fixing bacteria (like *Azotobacter chrococcum*), phosphate-soulbilizing bacteria (such as *Bacillus megaterium*), and potassium-solubilizing bacteria (like *B. mucilaginosus*) can be added to top layer of soil. Use of Arbuscular Mycorrhizal fungus as allied colonizers and biofertilizers helps for plant growth and play a critical role in ecosystem [115];
3. Application of seeds coated with suitable rhizobacteria in disrupted soil to help them stabilize at the root in order to arbitrate an effective bioremediation method [116];
4. Use of recombinant or GM bacterial strains possessing active biological containment trait that helps them to multiply and remove pollutant [117].

To achieve site specific microbial adaptation, it is crucial to practice indigenous mycorrhizal fungi and bacterial strains that are highly adapted to the initial climatic and soil conditions of the site. Also, the suitable host plant may help to stimulate mycorrhizal propagules present in the topsoil [114, 118]. In passive treatment of acid mine drainage affected by pyrite and sulphur oxidation, *Thiobacillus* and *Ferroplasma* bacterial species is very useful [118]. The application of PGPR in the field promotes the reclamation of mine tailings, while the use of manure, compost and fertilizer reduces the natural grass species [119]. Adaptive mechanism such as mycorrhizal association with specific plant species have also increase their capability to survive in low available phosphate soils [120]. Especially, Ectomycorrhiza fungi can immobilize heavy metals in soil, so it dropping the accessibility of metals to plants, promoting the revegetation at heavy metal contaminated sites. Therefore, plants habitat of these fungi implies a significant tolerance to potentially toxic trace metals [58, 121].

Another additional strategy is also applied by mixing the biochar with innate legume species such as *Acacia* sp. and other soil amendments like manure, compost or lime helps to improve the soil nutrients and plant growth at mine site [122]. Detailed information on the taxonomic and phylogenetic relationships within the species of mycorrhizal fungi and bacteria have provided by the recently developed molecular techniques such as PCR single-strand conformation polymorphism (PCR-SSCP), terminal restriction fragment length polymorphism (T-RFLP) or Denaturing gradient gel electrophoresis (DGGE). These techniques can help to recognize co-evolution and host-specificity of plants and mycorrhizal fungi, as well as the nature of the mycorrhizal species in the field [123]. Progress towards deciphering influence of mycorrhizae and associated bacteria in revegetation at polluted sites is a challenge.

8 Role of Advance Molecular Techniques and Genetically Engineered Organism in Remediation of Mining Sites

Horizontal gene transfer (HGT) is a transfer of gene between the intra-species populations during evolution, which result into the microbial adaptation to complex environments like metal mines and organic contamination sites. These transferred genes are mostly converted and adaptive with local interactions and can be related with the selective ecosystem of the microbial community. The extremophiles red alga, *Galdieria sulphuraria*, is an example of HGT. In the case of this red algae, soil has shown flexibility in its development through various nutritional and heavy metal detoxification approaches, as genes responsible for such morphological traits must have been transferred horizontally from the bacteria and Archaea to the algae living in the same environment [124]. Formation of a high level of ammonia, enhanced glutamate decarboxylase activity, surplus shock proteins production, defensive biological molecule production, and biofilm formation are the few key features of acid resistance mechanisms, adapted by tolerant bacteria [125]. A current report demonstrates that more than 20% part of bacterial genes and 40% part of archaeal genes are shifted horizontally to contribute in the microbial genome for evolution and adaptation in extreme environment conditions. And reproduction processes like conjugation, transduction and transformation contributes the main role in transferring the responsible proteins, either adaptive or immune to particular ecological habitat. A reverse gyrase protein found in hyperthermophilic feature has been transferred from an Archaea to bacteria for thermo-adaptation is the common example [126, 127]. Another studies on evolutionary process showed that adaptive trait responsible for resistance in *Thermus* sp. and *Deinococcus* sp., both belonging to relatively similar clad with distinct phenotypes, i.e. thermos-tolerant and radiation-tolerant, were initiated in an unified homologous megaplasmid, with certain genes either relocated or lost from other bacteria and Archaea [128]. This reveals that microbial community in neglected mine soil that must have altered to the complex environments like heavy metal tolerance, soil pH, temperature due to transfer of genes horizontally from intra species populations in due course of time. Mendez and Maier [60] studied revegetation in abandoned mine for period of 20 years and demonstrated that microbial and plant diversity was low as associated to a nearby uninterrupted location. This highlights the importance of field studies to authenticate the process of remediation in mine site ecosystems.

Recombinant DNA technology can be employed to introduce some aggressive strains that show tolerance for metals for improved microbe-assisted phytoremediation to form microbial communities in soil at abandoned mine sites [129, 130]. However, the survival rate and nutritional requirement of the recombinant microbes are the limiting factor related to genetic modifications [113, 131]. Some other constraints in the use of recombinant microbes involve its incapability to participate in the naturally occurring microbial community and bioavailability of varied toxic contaminants in soil. However, the strategies concerning plant-microbe symbiosis has found to resolve all these difficulties [113, 132]. The availability of appropriate plant species for symbiosis is very critical for effective removal of toxic metals [132].

9 Technological Limitations and Future Insight

Ecological factors like plant-soil and plant-microbe interaction systems performance a main part in refining the affected sites after the post mining process. Thus, manipulation of ecological inputs like application of the plant-microbe mixture for integrated plant-microbial remediation is the critical aspect for sustainable restoration of mining fields. But there is limited research on plant-microbe interactions and inter-genetic mechanism from derelict mines. This opens the way for widespread research on the characterization of the microbial groups present in mine areas [133]. In this context, the rhizosphere and phyllosphere habitats of prevalent plants will be quite significant. Further significant feature is the recognition of various classes of natural vegetation, such as persistent grasses, shrubs and certain native plants that can be grown in the mined areas. Considering the source of depletion and the resistance of metalliferous plants during restoration is indeed necessary for their improved endurance. To confirm the soil-plant systems that include the innate plants on human affected soils after manipulation of mine land, plant enhancement and plant community refurbishment is also crucial. This data will help to confer expansion of native established plant species in pre- and post-manipulated mine fields in subsequent stages [109]. The ultimate goal should be to achieve constant and purposeful plant communities in post mining environment with active microbiome.

10 Conclusion

The plant microbiome is important aspect to understand plant growth and its productivity, thus type of vegetation at polluted field is determined by the microbial community allied with mine soils and is found to be a critical aspect in the recovery process. In this context, fundamental understanding of biochemical signaling, translocation, and enablement between microbial community and associated plant is highly serious. Today's modern approaches such as meta-proteomic, meta-transcriptomic and meta-genomic can help to decipher complex mechanisms and interaction between microbiome and associated plants. A fully standardized meta-data on plant microbiome can surely recover the flora and increases the rate and extent of soil refinement. An important aspect to focus in mine site microbiology research is to manipulate the rhizosphere and non-rhizosphere microbiome to connect the potential of the microbial community for recovery of unrestricted mine sites and areas. An improved understanding of microbes variety at mine sites can open the way for substitute mining tactics like biomining, improved bioremediation approaches, supportable techniques in dispensation of the acidic mine drainages and removal of metals from wastewaters without any adverse outcomes on the surrounding environment.

References

1. Jaishankar M, Tseten T, Anbalagan N et al (2014) Toxicity, mechanism and health effects of some heavy metals. Interdisciplinary Toxicology 7(2):60–72. Slovak Toxicology Society. https://doi.org/10.2478/intox-2014-0009
2. Manhart A, Vogt R, Priester M, Dehoust G et al (2019) The environmental criticality of primary raw materials – A new methodology to assess global environmental hazard potentials of minerals and metals from mining. Miner Econ 32(1):91–107. https://doi.org/10.1007/s13563-018-0160-0
3. Claudio B (2011) Environmental impact of abandoned mine waste: a review. Nova Science Publishers Inc, New York
4. Rani N, Sharma HR, Kaushik A, Sagar A (2018) Bioremediation of mined waste land. in handbook of environmental materials management. Springer International Publishing, pp 1–25. https://doi.org/10.1007/978-3-319-58538-3_79-1
5. Mossa J, James LA (2013) 13.6 impacts of mining on geomorphic systems. Treat Geomorphol 13:74–95. https://doi.org/10.1016/B978-0-12-374739-6.00344-4
6. Tchounwou PB, Yedjou CG, Patlolla AK, Sutton DJ (2012) Heavy metal toxicity and the environment. EXS NIH Public Access 101:133–164. https://doi.org/10.1007/978-3-7643-8340-4_6
7. Wahsha M, Maleci L, Bini C (2018) The impact of former mining activity on soils and plants in the vicinity of an old mercury mine (Vallalta, belluno, ne italy). Geochem Explor Environ Anal 19(2):171–175. https://doi.org/10.1144/geochem2018-040
8. Fernandes CC, Kishi LT, Lopes EM et al (2018) Bacterial communities in mining soils and surrounding areas under regeneration process in a former ore mine. Brazilian J Microbiol 49(3):489–502. https://doi.org/10.1016/j.bjm.2017.12.006
9. Wei X, Wang X, Cao P, Gao Z et al (2020) Microbial community changes in the rhizosphere soil of healthy and rusty panax ginseng and discovery of pivotal fungal genera associated with rusty roots. BioMed Res Int. https://doi.org/10.1155/2020/8018525
10. Abatenh E, Gizaw B, Tsegaya Z, Wassie M (2017) Application of microorganisms in bioremediation-review. J Environ Microbiol 1(1):2–9
11. Tangahu BV, Sheikh Abdullah SR, Basri H et al (2011) A review on heavy metals (As, Pb, and Hg) uptake by plants through phytoremediation. Int J Chem Eng. https://doi.org/10.1155/2011/939161
12. Ojuederie OB, Babalola OO (2017) Microbial and plant-assisted bioremediation of heavy metal polluted environments: a review. Int J Environ Res Public Health 14(12). MDPI AG. https://doi.org/10.3390/ijerph14121504
13. Jafari M, Danesh YR, Goltapeh EM, Varma A (2013) Bioremediation and genetically modified organisms, pp 433–451. https://doi.org/10.1007/978-3-642-33811-3_19
14. Engwa GA, Ferdinand PU, Nwalo FN, Unachukwu MN (2019) Mechanism and health effects of heavy metal toxicity in humans. In: Karcioglu O, Arslan B (eds) Poisoning in the modern world—new tricks for an old dog? Intech Open. https://doi.org/10.5772/intechopen.82511
15. Godt J, Scheidig F, Grosse-Siestrup C, Esche V et al (2006) The toxicity of cadmium and resulting hazards for human health. J Occup Med Toxicol 1(1):22. https://doi.org/10.1186/1745-6673-1-22
16. Benavides MP, Gallego SM, Tomaro ML (2005) Cadmium toxicity in plants. Brazilian J Plant Physiol 17(1):21–34. Sociedade Brasileira de Fisiologia Vegetal. https://doi.org/10.1590/S1677-04202005000100003

17. González Á, Ayerbe L (2011) Response of coleoptiles to water deficit: growth, turgor maintenance and osmotic adjustment in barley plants (Hordeum vulgare L.). Agricul Sci 02(03):159–166. https://doi.org/10.4236/as.2011.23022
18. Gracia-Lor E, Sancho JV, Serrano R, Hernández F (2012) Occurrence and removal of pharmaceuticals in wastewater treatment plants at the Spanish mediterranean area of valencia. Chemosphere 87(5):453–462. https://doi.org/10.1016/j.chemosphere.2011.12.025
19. Oliveira H (2012) Chromium as an environmental pollutant: insights on induced plant toxicity. J Botany 2012:1–8
20. Straif K, Benbrahim-Tallaa L, Baan R et al (2009) A review of human carcinogens–part C: metals, arsenic, dusts, and fibres. Lancet Oncol 10(5):453–454. https://doi.org/10.1016/s1470-2045(09)70134-2
21. Sathawara NG, Parikh DJ, Agarwal YK (2004) Essential heavy metals in environmental samples from Western India. Bull Environ Contam Toxicol 73(4):756–761. https://doi.org/10.1007/s00128-004-0490-1
22. Bedi JS, Gill JPS, Aulakh RS et al (2013) Pesticide residues in human breast milk: risk assessment for infants from Punjab, India. Sci Total Environ 463–464:720–726. https://doi.org/10.1016/j.scitotenv.2013.06.066
23. Adriano DC, Adriano DC (2001) Introduction. In trace elements in terrestrial environments. Springer, New York, pp 1–27. https://doi.org/10.1007/978-0-387-21510-5_1
24. Cornu JY, Huguenot D, Jézéquel K, Lollier M, Lebeau T (2017) Bioremediation of copper-contaminated soils by bacteria. World J Microbiol Biotechnol 33:26. https://doi.org/10.1007/s11274-016-2191-4
25. Lehtola MJ, Miettinen IT, Keinänen MM et al (2004) Microbiology, chemistry and biofilm development in a pilot drinking water distribution system with copper and plastic pipes. Water Res 38(17):3769–3779. https://doi.org/10.1016/j.watres.2004.06.024
26. Comyns AE (2000) Handbook of copper compounds and applications. In: Richardson HW (ed) Marcel dekker, New York. ISBN 0-8247-8998-9. Appl Organometall Chem 14(3):174–175. https://doi.org/10.1002/(sici)1099-0739(200003)14:3%3c174::aid-aoc940%3e3.0.co;2-g
27. Van Hullebusch E, Chatenet P, Deluchat V, Chazal PM et al (2003) Copper accumulation in a reservoir ecosystem following copper sulfate treatment (St. Germain Les Belles, France). Water Air Soil Pollut 150(1–4):3–22. https://doi.org/10.1023/A:1026148914108
28. Meyer JS, Boese CJ, Morris JM (2007) Use of the biotic ligand model to predict pulse-exposure toxicity of copper to fathead minnows (*Pimephales promelas*). Aquatic Toxicol 84(2):268–278. https://doi.org/10.1016/j.aquatox.2006.12.022
29. Tang D, Li TY, Liu JJ, Zhou ZJ et al (2008) Effects of prenatal exposure to coal-burning pollutants on children's development in China. Environ Health Perspect 116(5):674–679. https://doi.org/10.1289/ehp.10471
30. Steenland K, Boffetta P (2000) Lead and cancer in humans: where are we now? Am J Ind Med 38(3):295–299. https://doi.org/10.1002/1097-0274(200009)38:3%3c295:AID-AJIM8%3e3.0.CO;2-L
31. Mäki-Paakkanen J, Sorsa M, Vainio H (2009) Chromosome aberrations and sister chromatid exchanges in lead-exposed workers. Hereditas 94(2):269–275. https://doi.org/10.1111/j.1601-5223.1981.tb01764.x
32. Sharma P, Dubey RS (2005) Lead toxicity in plants. Brazilian J Plant Physiol. Sociedade Brasileira de Fisiologia Vegetal. 17(1):35–52. https://doi.org/10.1590/s1677-04202005000100004
33. Cersosimo MG, Koller WC (2006) The diagnosis of manganese-induced parkinsonism. Neuro Toxicol 27(3):340–346. https://doi.org/10.1016/j.neuro.2005.10.006
34. Cowan DM, Zheng W, Zou Y, Shi X et al (2009) Manganese exposure among smelting workers: relationship between blood manganese-iron ratio and early onset neurobehavioral alterations. Neuro Toxicol 30(6):1214–1222. https://doi.org/10.1016/j.neuro.2009.02.005
35. Barbara G, Wojciech D, Aneta B, Paulina B, Olga B, Justyna W (2017) air contamination by mercury, emissions and transformations—a review. Water Air Soil Pollut 228(4):1–31. https://doi.org/10.1007/s11270-017-3311-y

36. Antunes dos Santos A, Appel Hort M, Culbreth M et al (2016) Methylmercury and brain development: A review of recent literature. J Trace Elem Med Biol 38:99–107. https://doi.org/10.1016/j.jtemb.2016.03.001
37. Filella M, Belzile N, Chen YW (2002) Antimony in the environment: A review focused on natural waters I Occurence. Earth Sci Rev 57(1–2):125–176. https://doi.org/10.1016/S0012-8252(01)00070-8
38. Filella M, Williams PA, Belzile N (2009) Antimony in the environment: knowns and unknowns. Environ Chem 6:95–105. https://doi.org/10.1071/EN09007_AC
39. Li J, Wang Q, Oremland RS, Kulp TR et al (2016) Microbial antimony biogeochemistry: enzymes, regulation, and related metabolic pathways. Appl Environ Microbiol Am Soc Microbiol 82(18):5482–5495. https://doi.org/10.1128/AEM.01375-16
40. Tschan M, Robinson BH, Nodari M, Schulin R (2009) Antimony uptake by different plant species from nutrient solution, agar and soil. Environ Chem 6:144–152. https://doi.org/10.1071/EN08103
41. Sundar S, Chakravarty J (2010) Antimony toxicity. Int J Environ Res Public Health. Multidisciplinary Digital Publishing Institute (MDPI) 7(12):4267–4277. https://doi.org/10.3390/ijerph7124267
42. Wang S, Mulligan CN (2006) Occurrence of arsenic contamination in Canada: Sources, behavior and distribution. Sci Total Environ 366(2–3):701–721. https://doi.org/10.1016/j.scitotenv.2005.09.005
43. Ratnaike RN (2003) Acute and chronic arsenic toxicity. Postgrad Med J Fellowship Postgrad Med 79(933):391–396 https://doi.org/10.1136/pmj.79.933.391
44. Stayner L, Welch LS, Lemen R (2013) The worldwide pandemic of asbestos-related diseases. Annual Rev Public Health 34(1):205–216. https://doi.org/10.1146/annurev-publhealth-031811-124704
45. Walters GI, Robertson AS, Bhomra PS, Burge PS (2018) Asbestosis is prevalent in a variety of construction industry trades. NPJ Primary Care Respirat Med 28(1):11. https://doi.org/10.1038/s41533-018-0078-6
46. Bhandari J, Thada PK, Sedhai YR (2020) Asbestosis. Stat Pearls Publishing, In StatPearls
47. Lotti M, Bergamo L, Murer B (2010) Occupational toxicology of asbestos-related malignancies. Clinical Toxicol (Philadelphia, Pa.) 48(6):485–496. https://doi.org/10.3109/15563650.2010.506876
48. Liao M, Xie XM (2007) Effect of heavy metals on substrate utilization pattern, biomass, and activity of microbial communities in a reclaimed mining wasteland of red soil area. Ecotoxicol Environ Safety 66(2):217–223. https://doi.org/10.1016/j.ecoenv.2005.12.013
49. Hur M, Kim Y, Song HR, Kim JM et al (2011) Effect of genetically modified poplars on soil microbial communities during the phytoremediation of waste mine tailings. Appl Environ Microbiol 77:7611–7619. https://doi.org/10.1515/biolog-2017-0117
50. Gołebiewski M, Deja-Sikora E, Cichosz M, Tretyn A, Wróbel B (2014) 16S rDNA pyrosequencing analysis of bacterial community in heavy metals polluted soils. Microbial Ecol 67(3):635–647. https://doi.org/10.1007/s00248-013-0344-7
51. Alvarez A, Saez JM, Davila Costa JS, Colin VL et al (2017) Actinobacteria: current research and perspectives for bioremediation of pesticides and heavy metals. Chemosphere. Elsevier Ltd. https://doi.org/10.1016/j.chemosphere.2016.09.070
52. Thavamani P, Samkumar RA, Satheesh V et al (2017) Microbes from mined sites: Harnessing their potential for reclamation of derelict mine sites. Environ Pollut (Barking, Essex: 1987) 230:495–505. https://doi.org/10.1016/j.envpol.2017.06.056
53. Xie P, Hao X, Herzberg M, Luo Y et al (2015) Genomic analyses of metal resistance genes in three plant growth promoting bacteria of legume plants in Northwest mine tailings, China. J Environ Sci (China) 27(C):179–187. https://doi.org/10.1016/j.jes.2014.07.017
54. Chen L, Hu M, Huang L et al (2015) Comparative metagenomic and metatranscriptomic analyses of microbial communities in acid mine drainage. ISME J 9:1579–1592. https://doi.org/10.1038/ismej.2014.245

55. Devasia P, Natarajan KA (2004) Bacterial Leaching. Biotechnology in the mining industry, resonance, pp 27–34
56. Goldstein A, Lester T, Brown J (2003) Research on the metabolic engineering of the direct oxidation pathway for extraction of phosphate from ore has generated preliminary evidence for PQQ biosynthesis in Escherichia coli as well as a possible role for the highly conserved region of quinoprotein dehydrogenases. Biochimica etBiophysica Acta Proteins Proteom 1647:266–271. https://doi.org/10.1016/S1570-9639(03)00067-0
57. García-Moyano A, González-Toril E, Aguilera Á, Amils R (2012) Comparative microbial ecology study of the sediments and the water column of the Río Tinto, an extreme acidic environment. FEMS Microbiol Ecol 81(2):303–314. https://doi.org/10.1111/j.1574-6941.2012.01346.x
58. Brown SD, Utturkar SM, Klingeman DM et al (2012) Twenty-one genome sequences from pseudomonas species and 19 genome sequences from diverse bacteria isolated from the rhizosphere and endosphere of *Populus deltoides*. J Bacteriol. https://doi.org/10.1128/JB.01243-12
59. Xie X, Xiao S, He Z et al (2007) Microbial populations in acid mineral bioleaching systems of tong shankou copper mine China. J Appl Microbiol 103(4):1227–1238. https://doi.org/10.1111/j.1365-2672.2007.03382.x
60. Mendez MO, Maier RM (2008) Phytostabilization of mine tailings in arid and semiarid environments–an emerging remediation technology. Environ Health Perspect 116(3):278–283. https://doi.org/10.1289/ehp.10608
61. Kay CM, Rowe OF, Rocchetti L, Coupland K et al (2013) Evolution of microbial "Streamer" growths in an acidic, metal-contaminated stream draining an abandoned underground copper mine. Life 3(1):189–210. https://doi.org/10.3390/life3010189
62. Johnson DB, Hallberg KB, Hedrich S (2014) Uncovering a microbial enigma: isolation and characterization of the streamer-generating, iron-oxidizing, acidophilic bacterium "Ferrovum myxofaciens". Appl Environ Microbiol 80(2):672–680. https://doi.org/10.1128/AEM.03230-13
63. Wielinga B, Lucy JK, Moore JN et al (1999) Microbiological and geochemical characterization of fluvially deposited sulfidic mine tailings. Appl Environ Microbiol 65(4):1548–1555. https://doi.org/10.1128/aem.65.4.1548-1555.1999
64. Santofimia E, González-Toril E, López-Pamo E et al (2013) Microbial diversity and its relationship to physicochemical characteristics of the water in two extreme acidic pit lakes from the iberian pyrite belt (SW Spain). PLoS ONE 8(6): https://doi.org/10.1371/journal.pone.0066746
65. Reiche M, Lu S, Ciobotă V, Neu TR et al (2011) Pelagic boundary conditions affect the biological formation of iron-rich particles (iron snow) and their microbial communities. Limnol Oceanogr 56(4):1386–1398. https://doi.org/10.4319/lo.2011.56.4.1386
66. Navarro-Noya YE, Jan-Roblero J, del Carmen González-Chávez M et al (2010) Bacterial communities associated with the rhizosphere of pioneer plants (*Bahia xylopoda* and *Viguiera linearis*) growing on heavy metals-contaminated soils. Antonie Van Leeuwenhoek 97:335–349. https://doi.org/10.1007/s10482-010-9413-9
67. Kent AD, Triplett EW (2002) Microbial communities and their interactions in soil and rhizosphere ecosystems. Annual Rev Microbiol 56:211–236. https://doi.org/10.1146/annurev.micro.56.012302.161120
68. Vorholt JA (2012) Microbial life in the phyllosphere. Nature Rev Microbiol 10(12):828–840. https://doi.org/10.1038/nrmicro2910
69. Sánchez-López AS, Del Carmen A, González-Chávez M, Carrillo-González R et al (2015) Wild flora of mine tailings: perspectives for use in phytoremediation of potentially toxic elements in a semi-arid region in Mexico. Int J Phytoremed 17(1–6):476–484. https://doi.org/10.1080/15226514.2014.922922
70. Morgan JAW, Bending GD, White PJ (2005) Biological costs and benefits to plant-microbe interactions in the rhizosphere. J Experim Botany 56:1729–1739. https://doi.org/10.1093/jxb/eri205

71. Batty LC (2005) The potential importance of mine sites for biodiversity. Mine Water Environ 24(2):101–103. https://doi.org/10.1007/s10230-005-0076-0
72. Khan AG (2005) Role of soil microbes in the rhizospheres of plants growing on trace metal contaminated soils in phytoremediation. J Trace Elements Med Biol Elsevier GmbH 18:355–364. https://doi.org/10.1016/j.jtemb.2005.02.006
73. Hayat R, Ali S, Amara U, Khalid R, Ahmed I (2010) Soil beneficial bacteria and their role in plant growth promotion: a review. Annals Microbiol. https://doi.org/10.1007/s13213-010-0117-1
74. Shetty KG, Hetrick BAD, Figge DAH, Schwab AP (1994) Effects of mycorrhizae and other soil microbes on revegetation of heavy metal contaminated mine spoil. Environ Pollut 86(2):181–188. https://doi.org/10.1016/0269-7491(94)90189-9
75. Baker BJ, Hugenholtz P, Dawson SC, Banfield JF (2003) Extremely acidophilic protists from acid mine drainage host Rickettsiales-lineage endosymbionts that have intervening sequences in their 16S rRNA genes. Appl Environ Microbiol 69(9):5512–5518. https://doi.org/10.1128/aem.69.9.5512-5518.2003
76. Ghnaya T, Mnassri M, Ghabriche R, Wali M et al (2015) Nodulation by S*inorhizobiummeliloti* originated from a mining soil alleviates Cd toxicity and increases Cd-phytoextraction in *Medicagosativa* L. Front Plant Sci 6:863. https://doi.org/10.3389/fpls.2015.00863
77. Hrynkiewicz K, Baum C, Niedojadło J, Dahm H (2009) Promotion of mycorrhiza formation and growth of willows by the bacterial strain *Sphingomonas* sp. 23L on fly ash. Biol Fertil Soils 45:385–394. https://doi.org/10.1007/s00374-008-0346-7
78. Suharno S, Soetarto ES, Sancayaningsih RP, Kasiamdari RS (2017) Association of arbuscular mycorrhizal fungi (AMF) with *Brachiaria precumbens* (Poaceae) in tailing and its potential to increase the growth of maize (Zea mays). Biodiversitas J Biol Diver 18(1). https://doi.org/10.13057/biodiv/d180157
79. Mahar A, Wang P, Ali A, Awasthi MK et al (2016) Challenges and opportunities in the phytoremediation of heavy metals contaminated soils: a review. Ecotoxicol Environ Safety 126:111–121. https://doi.org/10.1016/j.ecoenv.2015.12.023
80. Cervantes C, Campos-García J, Devars S et al (2001) Interactions of chromium with microorganisms and plants. FEMS Microbiol Rev 25(3):335–347. https://doi.org/10.1111/j.1574-6976.2001.tb00581.x
81. Arriagada CA, Herrera MA, Borie F et al (2007) Contribution of arbuscular mycorrhizal and saprobe fungi to the aluminum resistance of *Eucalyptus globulus*. Water Air Soil Pollut 182:383–394. https://doi.org/10.1007/s11270-007-9349-5
82. Liang L, Liu W, Sun Y et al (2017) Phytoremediation of heavy metal contaminated saline soils using halophytes: current progress and future perspectives. Environ Rev 25:269–281. https://doi.org/10.1139/er-2016-0063
83. Sheoran V, Sheoran AS, Poonia P (2016) Factors affecting phytoextraction: a review. Pedosphere 26(2):148–166. https://doi.org/10.1016/S1002-0160(15)60032-7
84. Magdziak Z, Gąsecka M, Goliński P, Mleczek M (2015) Phytoremediation and environmental factors. Phytoremed Manag Environ Contam 1:45–56. Springer International Publishing. https://doi.org/10.1007/978-3-319-10395-2_4
85. Luo J, Qi S, Gu XWS et al (2016) An evaluation of EDTA additions for improving the phytoremediation efficiency of different plants under various cultivation systems. Ecotoxicology 25(4):646–654. https://doi.org/10.1007/s10646-016-1623-0
86. Kazemalilou S, Delangiz N, AsgariLajayer B, Ghorbanpour M (2020) Insight into plant-bacteria-fungi interactions to improve plant performance via remediation of heavy metals: an overview. Molecul Aspects Plant Benef Microbes Agricul Elsevier 123–132. https://doi.org/10.1016/b978-0-12-818469-1.00010-9
87. Saffari VR, Saffari M (2020) Effects of EDTA, citric acid, and tartaric acid application on growth, phytoremediation potential, and antioxidant response of *Calendula officinalis* L. in a cadmium-spiked calcareous soil. Int J Phytoremed 22(11):1204–1214. https://doi.org/10.1080/15226514.2020.1754758

88. Chen H, Cutright T (2001) EDTA and HEDTA effects on Cd, Cr, and Ni uptake by *Helianthus annuus*. Chemosphere 45(1):21–28. https://doi.org/10.1016/S0045-6535(01)00031-5
89. Wei SH, Zhou QX, Wang X et al (2005) A newly found Cd-hyperaccumulator *Solanum nigrum* L. Chinese Bulletin Sci 50(1):33–38. https://www.doi.org/10.1360/982004-292
90. Wu J, Chen A, Peng S, Wei Z, Liu G (2013) Identification and application of amino acids as chelators in phytoremediation of rare earth elements lanthanum and yttrium. Plant Soil 373(1–2):329–338. https://doi.org/10.1007/s11104-013-1811-0
91. Jagetiya B, Sharma A (2013) Optimization of chelators to enhance uranium uptake from tailings for phytoremediation. Chemosphere 91(5):692–696. https://doi.org/10.1016/j.chemosphere.2012.11.044
92. Singh S, Fulzele DP, Kaushik CP (2016) Potential of *Vetiveriazizanoides* L. Nash for phytoremediation of plutonium (^{239}Pu): chelate assisted uptake and translocation. Ecotoxicol Environ Safety 132:140–144. https://doi.org/10.1016/j.ecoenv.2016.05.006
93. Wang K, Liu Y, Song Z, Wang D, Qiu W (2019) Chelator complexes enhanced *Amaranthus hypochondriacus* L. phytoremediation efficiency in Cd-contaminated soils. Chemosphere 237. https://doi.org/10.1016/j.chemosphere.2019.124480
94. MOSPI (Ministry of Statistics and Programme Implementation) (2013) Statistical year book India. Chapter−15, Mining, Published by Ministry of statistics and programme implementation, Government of India
95. Kamaludeen SP, Ramasamy K (2008) Rhizoremediation of metals: harnessing microbial communities. Indian J Microbiol 48(1):80–88. https://doi.org/10.1007/s12088-008-0008-3
96. Wood TK (2008) Molecular approaches in bioremediation. Current Opin Biotechnol. https://doi.org/10.1016/j.copbio.2008.10.003
97. Saravanan A, Jeevanantham S, Narayanan VA, Kumar PS et al (2020) Rhizoremediation−a promising tool for the removal of soil contaminants: a review. J Environ Chem Eng 8(2): https://doi.org/10.1016/j.jece.2019.103543
98. Doty SL (2008) Enhancing phytoremediation through the use of transgenics and endophytes. New Phytol 179(2):318–333. https://doi.org/10.1111/j.1469-8137.2008.02446.x
99. Ali H, Khan E, Sajad MA (2013) Phytoremediation of heavy metals-concepts and applications. Chemosphere 91(7):869–881. https://doi.org/10.1016/j.chemosphere.2013.01.075
100. Saxena G, Purchase D, Mulla SI et al (2020) Phytoremediation of heavy metal-contaminated sites: eco-environmental concerns, field studies, sustainability issues, and future prospects. Rev Environ Contam Toxicol 249:71–131. Springer, New York LLC. https://doi.org/10.1007/398_2019_24
101. Gupta G, Khan J, Singh NK (2020) Phytoremediation of metal-contaminated Sites. In: Hasanuzzaman M (ed) Plant ecophysiology and adaptation under climate change: mechanisms and perspectives II. Springer, pp 726–747. https://doi.org/10.1007/978-981-15-2172-0_27
102. Gomes MA da C, Hauser-Davis RA, de Souza AN, Vitória AP (2016) Metal phytoremediation: general strategies, genetically modified plants and applications in metal nanoparticle contamination. Ecotoxicol Environ Safety 134:133–147. https://doi.org/10.1016/j.ecoenv.2016.08.024
103. Muthusaravanan S, Sivarajasekar N, Vivek JS, Paramasivan T et al (2018) Phytoremediation of heavy metals: mechanisms, methods and enhancements. Environ Chem Lett 16:1339–1359. https://doi.org/10.1007/s10311-018-0762-3
104. Gil-Loaiza J, White SA, Root RA, Solís-Dominguez FA et al (2016) Phytostabilization of mine tailings using compost-assisted direct planting: translating greenhouse results to the field. Sci Total Environ 565:451–461. https://doi.org/10.1016/j.scitotenv.2016.04.168
105. Arya SS, Devi S, Angrish R, Singal I, Rani K (2017) Soil reclamation through phytoextraction and phytovolatilization. In: Volatiles and Food Security: Role of Volatiles in Agro-Ecosystems. Springer, Singapore, pp 25–43. https://doi.org/10.1007/978-981-10-5553-9_3
106. Rai PK (2008) Heavy metal pollution in aquatic ecosystems and its phytoremediation using wetland plants: an ecosustainable approach. Int J Phytoremed 10(2):131–158. https://doi.org/10.1080/15226510801913918

107. Zhuang X, Chen J, Shim H, Bai Z (2007) New advances in plant growth-promoting rhizobacteria for bioremediation. Environ Int 33(3):406–413. https://doi.org/10.1016/j.envint.2006.12.005
108. Abou-Shanab RA, Ghanem K, Ghanem N, Al-Kolaibe A (2008) The role of bacteria on heavy-metal extraction and uptake by plants growing on multi-metal-contaminated soils. World J Microbiol Biotechnol 24:253–262. https://doi.org/10.1007/s11274-007-9464-x
109. Guo J, Muhammad H, Lv X, Wei T et al (2020) Prospects and applications of plant growth promoting rhizobacteria to mitigate soil metal contamination: a review. Chemosphere 246: https://doi.org/10.1016/j.chemosphere.2020.12582
110. Bernoth L, Firth I, McAllister P, McAllister P, Rhodes S (2000) Biotechnologies for remediation and pollution control in the mining industry. Mining Metall Explor 17:105–111. https://doi.org/10.1007/BF03402836
111. Tyagi M, Da Fonseca MM, de Carvalho CC (2011) Bioaugmentation and biostimulation strategies to improve the effectiveness of bioremediation processes. Biodegradation 22(2):231–241. https://doi.org/10.1007/s10532-010-9394-4
112. Dixit R, Wasiullah Malaviya D, Pandiyan K, Singh UB et al (2015) Bioremediation of heavy metals from soil and aquatic environment: an overview of principles and criteria of fundamental processes. MDPI AG, Sustainability (Switzerland). https://doi.org/10.3390/su7022189
113. Megharaj M, Ramakrishnan B, Venkateswarlu K et al (2011) Bioremediation approaches for organic pollutants: A critical perspective. Environ Int. Elsevier Ltd. https://doi.org/10.1016/j.envint.2011.06.003
114. Lindemann WC, Lindsey DL, Fresquez PR (1984) Amendment of Mine Spoil to Increase the Number and Activity of Microorganisms. Soil Sci Soc Am J 48(3):574–578. https://doi.org/10.2136/sssaj1984.03615995004800030021x
115. Patowary K, Patowary R, Kalita MC, Deka S (2016) Development of an efficient bacterial consortium for the potential remediation of hydrocarbons from contaminated sites. Front Microbiol 7:1092. https://doi.org/10.3389/fmicb.2016.01092
116. Kuiper I, Lagendijk EL, Bloemberg GV, Lugtenberg BJJ (2004) Rhizoremediation: a beneficial plant-microbe interaction. Molecular Plant-Microbe Interactions. Am Phytopathol Soc. https://doi.org/10.1094/MPMI.2004.17.1.6
117. Ronchel MC, Ramos JL (2001) Dual system to reinforce biological containment of recombinant bacteria designed for Rhizoremediation. Appl Environ Microbiol 67(6):2649–2656. https://doi.org/10.1128/AEM.67.6.2649-2656.2001
118. Hedin RS, Watzlaf GR, Nairn RW (1994) Passive treatment of acid mine drainage with limestone. J Environ Quality 23(6):1338–1345. https://doi.org/10.2134/jeq1994.00472425002300060030x
119. Grandlic CJ, Mendez MO, Chorover J et al (2008) Plant growth-promoting bacteria for phytostabilization of mine tailings. Environ Sci Technol 42(6):2079–2084. https://doi.org/10.1021/es072013j
120. Dodd JC (2000) The role of arbuscular mycorrhizal fungi in agro- and natural ecosystems. Outlook Agricul 29(1):55–62. https://doi.org/10.5367/000000000101293059
121. Hassan A, Pariatamby A, Ahmed A et al (2019) Enhanced bioremediation of heavy metal contaminated landfill soil using filamentous fungi consortia: a demonstration of bioaugmentation potential. Water Air and Soil Pollut 230(9). https://doi.org/10.1007/s11270-019-4227-5
122. Rawat J, Saxena J, Sanwal P (2019) Biochar: a sustainable approach for improving plant growth and soil properties. In Biochar—an imperative amendment for soil and the environment. Intech Open. https://doi.org/10.5772/intechopen.82151
123. Read DJ (2003) Towards ecological relevance—progress and pitfalls in the path towards an understanding of mycorrhizal functions in nature, pp 3–29. https://doi.org/10.1007/978-3-540-38364-2_1
124. Schönknecht G, Chen WH, Ternes CM, Barbier GG et al (2013) Gene transfer from bacteria and archaea facilitated evolution of an extremophilic eukaryote. Science 339(6124):1207–1210. https://doi.org/10.1126/science.1231707

125. Liu Y, Tang H, Lin Z, Xu P (2015) Mechanisms of acid tolerance in bacteria and prospects in biotechnology and bioremediation. Biotechnol Adv. Elsevier Inc. https://doi.org/10.1016/j.biotechadv.2015.06.001
126. Averhoff B, Müller V (2010) Exploring research frontiers in microbiology: recent advances in halophilic and thermophilic extremophiles. Res Microbiol 161(6):506–514. https://doi.org/10.1016/j.resmic.2010.05.006
127. Polz MF, Alm EJ, Hanage WP (2013) Horizontal gene transfer and the evolution of bacterial and archaeal population structure. Trends Genet TIG 29(3):170–175. https://doi.org/10.1016/j.tig.2012.12.006
128. Omelchenko MV, Wolf YI, Gaidamakova EK et al (2005) Comparative genomics of thermus thermophilus and deinococcus radiodurans: divergent routes of adaptation to thermophily and radiation resistance. BMC Evol Biol 5. https://doi.org/10.1186/1471-2148-5-57
129. Gamalero E, Lingua G, Berta G, Glick BR (2009) Beneficial role of plant growth promoting bacteria and arbuscular mycorrhizal fungi on plant responses to heavy metal stress. Canadian J Microbiol 55(5):501–514. https://doi.org/10.1139/w09-010
130. Evangelou MWH, Robinson BH, Günthardt-Goerg MS, Schulin R (2012) Metal uptake and allocation in trees grown on contaminated land: implications for biomass production. Int J Phytoremed 15(1):77–90. https://doi.org/10.1080/15226514.2012.670317
131. Ramakrishnan B, Megharaj M, Venkateswarlu K, Sethunathan N, Naidu R (2011) Mixtures of environmental pollutants: Effects on microorganisms and their activities in soils. Rev Environ Contam Toxicol 211:63–120. https://doi.org/10.1007/978-1-4419-8011-3_3
132. Gawronski SW, Greger M, Gawronska H (2011) Plant taxonomy and metal phytoremediation. In: Sherameti I, Varma A (eds) Detoxification of heavy metals. Soil Biology, vol 30. Springer, Berlin, Heidelberg. https://doi.org/10.1007/978-3-642-21408-0_5
133. Ayangbenro AS, Babalola OO (2020) Reclamation of arid and semi-arid soils: The role of plant growth-promoting archaea and bacteria. Current Plant Biol. Elsevier B.V. https://doi.org/10.1016/j.cpb.2020.100173

Reduction of Hexavalent Chromium Using Microbial Remediation: A Case Study of Pauni and Taka Chromite Mines, Central India

Shweta V. Deote, A. B. Ingle, Swapnil Magar, and Ruchika Jain

Abstract Chromium is a naturally occurring heavy metal, valued for its resistance to corrosion, oxidation, and enhancement of hardenability. Such qualities make it important for various industries, especially steel making. Discharge of chromium from industrial effluent is one of the major causes of chromium contamination in the environment which is mainly present in its hexavalent state. Hexavalent chromium is highly soluble in nature and its carcinogenicity and mutagenicity make it highly toxic for humans, animals, plants, and microorganisms. Elevated levels of chromium in the environment inhibit most of the microorganisms, but also promote the selection of resistant species. Microbes belonging to such metal enriched area show discrete characteristics as well as discrete mechanism for dealing with such toxic conditions. Studies have shown that these microbes have unique quality of reducing chromium toxicity. Biotransformation of Cr (VI) to Cr (III) using bacteria is the most practically efficient approach. Ability of the bacteria to tolerate the chromium and to reduce it further into more tolerable species makes it very useful agent for remediation of the chromium contaminated soils. Important bacterial genus reported till date include, *Pseudomonas spp*, *Enterobacter spp*, *Escherichia coli*, *Ochrobactrum sp. Lysinibacillus spp*, *Acinetobacter spp. Microbacterium spp*, *Rhodococcus spp*, and *Bacillus spp.* We conducted a study in the chromium-rich soils of small chromite mines located at Taka in Nagpur and Pauni in Bhandara districts of Maharashtra in Central India. The aim of this study was the isolation and identification of chromium tolerant bacterial species having the potential for bioremediation. Our study has indicated that the isolated bacterial species, *Lysinibacillus macroides* (LC183868), *Acinetobactor pittii* (LC155825) and *Bacillus safensis* (LC155823), can tolerate high chromium concentration and showed promising results when tested for chromium reduction ability. This study is replicable and holds great promise for bioremediation of chromium-contaminated soils.

S. V. Deote (✉) · A. B. Ingle · S. Magar
Department of Microbiology, Seth Kesarimal Porwal College, Kamptee, Dist., Nagpur 441001, Maharashtra, India
e-mail: shweta_vdi@rediffmail.com

R. Jain
Naional Institute of Miner's Health, Nagpur, India

K. Randive et al. (eds.), *Innovations in Sustainable Mining*, Earth and Environmental Sciences Library, https://doi.org/10.1007/978-3-030-73796-2_11

Keywords Hexavalent chromium · *Lysinibacillus macroides* · *Acinetobacter pittii* · *Bacillus safensis* · Bioremediation · Biotransformation · Tolerance

1 Introduction

Chromium is a naturally occurring heavy metal that presents in the earth's crust, with different oxidation states ranging from chromium (III) to chromium (VI) but stable in the trivalent Cr (III) and Hexavalent form. Chromium is present in mines as chromite ore. Chromite ($FeCr_2O_4$) is important because it is the only economic ore of chromium, an essential element for a wide variety of metals, chemicals, and manufactured products. Many other minerals contain chromium, but none of them are found in deposits that can be economically mined to produce chromium [1].

Chromium recognized as heavy metal finds broad applications in industries of chromate manufacturing, electroplating, in the leather industry, dyes and pigment fabrication and wood preservation [2]. In nature, chromium occurs as chromite. In trace amount, Cr (III) required as an essential nutrient by plants, animals to carry out various metabolic activities [3]. In an animals like cattle role of chromium reported to enhance insulin action [4]. In contrast available Cr (VI) proving to be a pollutant and brings about mutagenic as well as carcinogenic effects [5].

2 Health Hazards Due to Chromium Toxicity

Naturally chromium is present in rocks and soil but the industrial activities has aggravated its concentration. It is important to note that chromium gets saturated in the environment mostly because of the increasing industrial use of chromium. The polluting nature of chromium increases further since chromium showcase nature of water solubility by which it possesses the feature of getting retained in biotic organisms for extended period and water also provide easy mobility from one place to another. This feature severely affects the life cycle of human, plants and animals as well as a microbial habitat. Chromium toxicity and their health hazards depend on its oxidation states. Exposure of chromium (VI) trioxide can damage nasal mucosa and perforation of the nasal septum, whereas exposure of insoluble (VI) compounds results in damage of lower respiratory tract [6]. Inhalation, ingestion, and dermal absorption are the major routes of entry of chromium in the human and animal body. Exposure of chromium compounds via air, water leads to asthma, nasal irritation and ulceration, skin irritation and allergies, dermatitis, eardrum perforation and lung carcinoma [7–9]. Several studies also showed carcinogenic potential of chromium (VI) [10].

3 Mode of Chromium Contamination of Soils

Chromium compounds are found in the environment due to the erosion of chromium-containing rocks and can be distributed by volcanic eruptions. In the aqueous system, chromium occurs in both trivalent and hexavalent forms.

Chromium is valued for its incredible resistance to corrosion, oxidation, wear & galling and enhancement of hardenability. It is widely used in the manufacturing of alloys along with other metals like nickel, cobalt, molybdenum, copper, titanium, zirconium, and vanadium and in the industries like Metallurgy, Chromium electroplating, Pigments and dyes, Wood preservatives and Tanning [11, 12].

Chromium enters into the environment (Soil, Water, and Air) through natural as well as anthropogenic sources. Hexavalent chromium is released as a by-product by several industrial activities. Chromium levels in soil vary according to the area and the degree of contamination. Mining activities near chromite mines, lead to deposition of a huge amount of material as well as accumulation of mine seepage waters which may be the main sources of chromium pollution of inland freshwater and farmlands in the vicinity of the mining sites [13, 14].

4 Effect of Hexavalent Chromium in Soils

4.1 Vegetation and Crops

Cr (VI) shows toxic effects on plants by affecting their growth and development at different stages. Chromium toxicity to plants has been well studied by Shankar et al. [15]. It gets absorbed by plants through polluted water and soil. Chromium accumulation inhibits plant growth, reduces pigment content and induces chlorosis in young leaves. It also affects seed germination; damages root cells and can cause inhibition of photosynthesis by altering enzymatic functions. [16, 17] In aquatic plants, its accumulation resulted in a decrease in biomass content [18, 19].

Accumulation of heavy metals in plants depends upon plant species, soil properties, and the efficiency of different plants [20]. Effect of chromium toxicity on plant growth has been studied on various plants e.g., Pea, rice, wheat, Maize, sunflower. Uptake of hexavalent and trivalent chromium has been studied in barley plant [21–23]. In *Nymphaea alba* L (European White water lily) chromium toxicity resulted in a reduction of chlorophyll biosynthesis, nitrate reductase activity, and protein content. [24]. Immobility of trivalent chromium limits its accumulation in the food chain; on the other hand, hexavalent chromium is a matter of concern [25].

4.2 Animal and Human Health

In humans, the respiratory tract is the major target site of exposure for chromium & its compounds. Occupational chromium exposure may induce DNA damage and can lead to lung cancer and other work-related diseases like asthma and nasal septum ulcers. [26–28] Toxicity of Cr (VI) is more due to its high membrane permeability and can cause a functional alteration in lungs, liver, pancreas, respiratory tract and kidney [29]. Hexavalent chromium can affect fetal development in mammals [30].

Various studies have been performed to determine the toxic effects of chromium in animals. In rats, exposure of Cr (VI) during the prepubertal stage of development may disturb normal testicular physiology at adulthood [31] also prolong exposure of hexavalent compounds can cause thickening of septa of the alveolar lumen, interstitial fibrosis bronchopneumonia, and lung abscesses. Nasal septum perforation, epithelial necrosis, and hyperplasia in the large and medium bronchi were observed in mice due to the chromium toxicity. It has been reported that at higher concentration, Cr (III) can reduce fertility in both male and female fertility [32].

4.3 Effect on Microorganisms

The soil is the main component of the environment for pollutants and trace elements discharged through various anthropogenic activities. Microorganisms are available in nature having ability to transform organic and inorganic materials along with many toxic elements, which are capable of altering the organism growth, morphology and metabolism. Microorganisms present in nature with prolonged exposure to such kind of metals build capability to tolerate metal concentration by genetic adaptation, which considered being stable. Microorganisms have developed various mechanisms of metal tolerance and resistance to deal with metals in the environment. Which include oxidation of metals, methylation, enzymatic reduction, metal-organic complexion, metal ligand degradation, metal efflux pumps, intracellular and extracellular metal sequestration, and exclusion by permeability barrier etc. [33–35]. Such isolates are termed as metal tolerant bacterial species having potential to biodegrade or bio accumulate toxic metal, which can ultimately reduce the load of metal in a given environment.

5 Bacterial Mechanisms of Chromium Tolerance

Increased levels of heavy metals like chromium are toxic to microbial habitat. Most of the microbes are sensitive to hexavalent chromium, but some species can tolerate its higher concentrations. Prolong exposure of heavy metals to microorganisms can develop resistant species. To deal with such toxic environments, microbes developed various mechanisms. Several microbes like bacteria, fungi, yeast; algae have been reported for their chromium resistance capacity [36–39].

To deal with chromium toxicity several bacterial species have been reported to evolved mechanisms like extracellular reduction of chromium from Cr (VI) to Cr (III), intracellular reduction, accumulation, biosorption, efflux mechanism, DNA repair mechanism, biotransformation, extrusion, use of enzymes, production of exopolysaccharide (EPS) etc. [40, 41]. Even though chromium is becoming a part of pollutants, but nature gives us some clues to control its polluting nature. Since soil represents its vast source of minerals, it also possesses the biological agents those can biotransform these compounds so that it will never reach to polluting level [42]. Microbial plasma membrane, cell wall, or capsule could prevent metal ions from entering the cell. Bacteria can adsorb metal ions by ionizable groups of the cell wall (amino, carboxyl, phosphate, and hydroxyl groups). Various environmental factors are responsible for biotransformation of metals like temperature, pH and redox potential and low molecular weight of organic acid [33].

In a success stories many bacterial species recorded with chromium tolerance capability belongs to gram-positive as well as gram-negative group such as *Pseudomonas spp* [43–45]; *Desulphovibrio spp* [46] *Enterobacter spp* [47]; *Escherichia coli* [48]; *Bacillus spp* [49]; *Ochrobactrum intermedium* [50]; *Lysinibacillus spp* [8, 51]; and *Acinetobacter spp.* Soil containing heavy metals are potential sources for identifying toxic metal-resistant bacteria as the presence of metals at sufficient concentrations resulted in metal tolerant species [52].

Based on above background the present study focusses on isolation of chromium tolerant microbes from chromite mine of Bhandara district and also study its Cr reduction ability. The isolate was also checked for biosafety usages regarding antibiotic sensitivity.

6 Methodology, Sampling, and Analysis

6.1 Sampling Method

To isolate chromium tolerant bacterial species, the soil from chromium rich mine soil of Pauni region with latitude and longitude 20°47'00" N, 79°39'00" E was collected as per standard protocol (Fig. 1) [53]. Once collected the physiochemical characterization of mine soil was done for mineral and other parameters such as chromium level, soil pH, electrical conductivity, organic carbon content, total nitrogen (Kjeldahl method), available phosphorus and potassium. The soil analysis (XRF) was out sourced from mineral testing lab, Indian Bureau of Mines, Nagpur. USEPA standard protocols were followed for collection of soil samples. Soil samples were stored and refrigerated for further processing.

Fig. 1 Chromite reservoir at Pauni and Taka

6.2 Isolation and Characterization of Cr-Reducing Bacteria

For selection of chromium tolerant bacterial species only, serially diluted soil samples inoculated as 100 μl on the Luria-Bertani (LB) medium evicted with 50 mg/L Cr in the form of potassium dichromate ($K_2Cr_2O_7$) salt. After inoculation the plates incubated at 37 °C for 48 h to record the colony appearance, colony morphology and isolate sub cultured on the same medium for maintenance.

6.3 Chromium Tolerance Test

To determine the Cr tolerance ability of selected isolates, the isolates were grown in different concentration of chromium (100–1200 mg/L) added in LB Broth. In LB medium chromium ($K_2Cr_2O_7$) concentration in the range 100–1300 mg/L in different groups and inoculated with culture with 1 O.D. of freshly grown culture. Once inoculated samples incubated for 96 h at 37 °C. During the course of time the microbial growth was estimated in every 24 h using spectrometer set at 600 nm.

6.4 Chromium Cr (VI) Reduction Assay

Cr (VI) is more toxic in comparison to Cr (III), thus Cr (VI) reduction ability addressed in the presence of promising isolate. The selected isolates were grown in the presence of LB Broth with Cr (VI) in the range (100–500 mg/L). The control test comprise of inoculum without Cr, keeping rest of the conditions same. All inoculated samples along with control incubated at 37 °C for 120 h. The response of reduction in chromate level recorded at an interval of 24 h up to 120 h. In a process, at every time interval, 10 ml of nutrient broth was centrifuged, and the supernatant was detected for Cr (VI) by 1, 5, -diphenyl carbazide by setting spectrophotometer at 540 nm using a spectrophotometer [35].

6.5 SEM and EDX Analysis

The ability of chromium tolerant bacterial species was structurally analyzed by involving SEM and EDX methodologies. In a process, cell samples treated with and without CP (VI) washed in a 0.1 M phosphate buffer saline. It was fixed overnight in 2% glutaraldehyde followed by rinsing in distilled water. It has then dehydrated under a series of ethanol concentrations (20, 40, 60, 80, 95 and 100%). Samples dried and kept in a desiccator until use. All samples then mounted on a brass-stub and sputter coated with gold and it transferred to scanning electron microscope coupled with Energy Dispersive X-ray spectrophotometer (EDX). The study was carried out in the Department of Metallurgy, Vishweshwarya National Institute of Technology, Nagpur, Maharashtra.

6.6 Antibiotic Susceptibility Test

Even though isolate from origin found to be useful for environmental studies; it is also essential to consider its antibiotic sensitivity profile. This test allows us to consider antibiogram for future control of this organism by updating ourselves with the sensitivity these organism showcases for a given antibiotic. Here the antibiotic sensitivity was carried out by Kirby-Bauer disc diffusion method [54, 55].

In a process cell, suspension turbidity maintained at 0.5 Mc Farland standards [56, 57]. Once bacteria inoculated on the medium, it does loaded with 24 antibiotics such as Amikacin (30 mcg), Augmentin (30 mcg), Ampicillin (10 mcg), Aztreonam (30 mcg), Cefoxitin (30 mcg), Ceftazidime (30 mcg), Ceftriaxone (30 mcg), Cefuroxime (30 mcg), Cephalothin (30 mcg), Chloramphenicol (30 mcg), Ciprofloxacin (5 mcg), Co-Trimoxazole (25 mcg), Fusidic Acid (30 mcg), Gentamicin (10 mcg), Imipenem (10 mcg), Linezolid (30 mcg), Meropenem (10 mcg), Norfloxacin (10 mcg), Ofloxacin (5 mcg), Piperacillin (100 mcg), Rifampicin (5 mcg), Tetracycline (30 mcg), Tobramycin (10 mcg), Vancomycin (30 mcg). All plates were recorded for the zone of inhibition when incubated at 37 °C for 24 h and formed the zone of inhibition was recorded in millimeter and compared with CLSI chart to ascertain exact sensitivity [56].

6.7 *Bacterial Identification*

Once the ability of isolates to reduce chromium has established, isolate was successfully identified by carrying out 16s rRNA gene sequencing. Here Genomic DNA was purified by the modified CTAB method. It was then targeted by 16S primer as 27 forward (AGAGTTTGATCMTGGCTCAG) and 1492 reverse (ACGGYTACCTTGTTACGACTT). The reaction volume for PCR set as 32.0 μl nuclease free water, 5.0 μl PCR buffer 10x, 2.0 μldNTP (10 mM), 4.0 μl forward primer (10 μM), 4.0 μl reverse primer (10 μM), 1.0 μl Taq DNA polymerase enzyme (1U/μl) and 200 ng DNA template. PCR reaction was programmed as, Initial denaturation of 3 min. at 94 °C, denaturation of 1 min. at 94 °C, primer annealing for 1 min. at 54 °C, extension of 2 min. at 72 °C, final extension for 5 min. at 72 °C; total 30 cycles and stored at 4 °C. The nucleotide sequences were compared to the NCBI GenBank using the nBLAST tool and was submitted to GenBank using Sequin.

7 Result and Discussion

7.1 *Soil Features*

The physiochemical characteristics of mine soil of Pauni was recorded with pH 7.2, Electric conductivity as 0.21 deci/m, organic carbon as 0.50%, Total Nitrogen (Kjeldhal Method) as 100 kg/ha, Potassium as 40.32 kg/ha. The XRF analysis of Pauni soils contains chromium about 23.82 g/Kg indicated the very high load of chromium in Pauni mine.

7.2 *Identification of Promising Isolates*

The organisms from the soil were isolated by serial dilution method and plated. Total 14 isolates were selected and inoculated on specific media rich in $K_2Cr_2O_7$ depending on different colony morphology. Once this isolation ($n = 20$) regularly sub-cultured on $K_2Cr_2O_7$ containing LB medium, isolated LB 1.1, NA + LB & M9 + LB were able to grow profoundly up to 900 even after many subcultures (Fig. 2).

7.3 *Evaluation of Chromium Tolerance*

Isolates LB 1.1, NA + LB & M9 + LB were capable of tolerating chromium in a medium analyzed at 100, 300, 500, 700, 900 and 1100 mg/L which result in varied response recorded as O.D. at 600 nm with 24 h interval of recording.

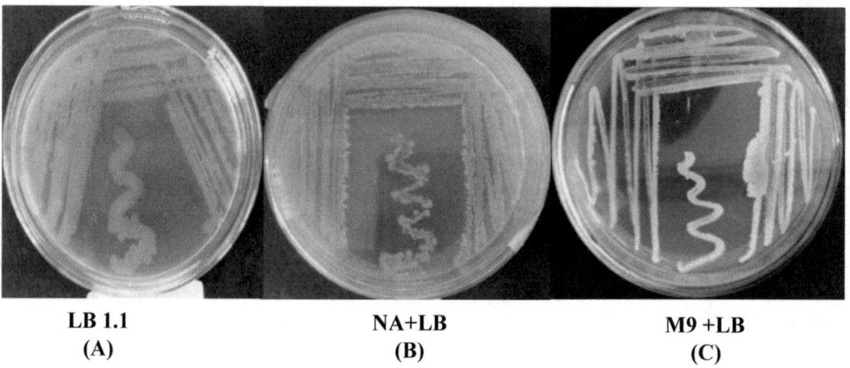

Fig. 2 Chromium tolerant isolates

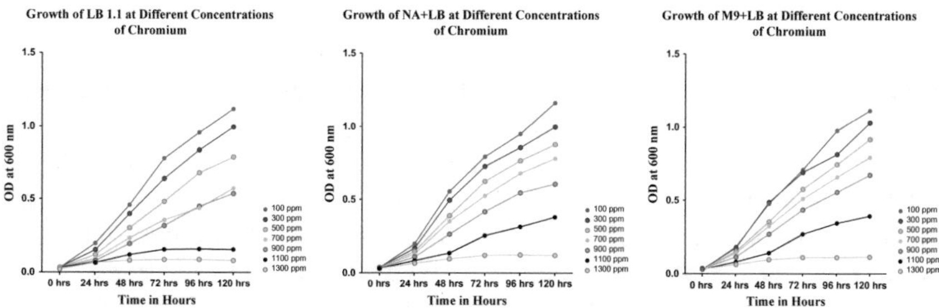

Fig. 3 Growth Pattern of LB 1.1, NA + LB & M9 + LB on different concentrations of Chromium

As per Fig. 3 isolate registered the expected high O.D. at the lower concentration, i.e., 100 mg/L to 120 h and as the chromium concentration went on increasing. The increased O.D. recorded regular fall of 0.20 O.D. on an average till 24 h. It has also recorded that once the chromium concentration reaches beyond 300 mg/L, a steep fall in O.D. recorded with isolate LB 1.1 as in Fig. 1. As per observation between 300–700 mg/L fall in activity remains to be around 0.37 O.D. ever higher in any setting. Here it could be said that up to 800 mg/L isolate LB 1.1 retained a strong increase in O.D. up to 0.36 O.D. at 660 nm, but with 1100 mg/L and 1200 mg/L meager growth was recorded as evidenced whereas NA + LB and M9 + LB shows satisfactory growth as compared to LB1.1 at 900 mg/L and 1100 mg/L. This shows that the LB 1.1 strain has tolerance limit up to 900 mg and NA + LB & LB 1.1 can tolerate up to 1200 mg/L concentration of Cr which is very high for any strain.

7.4 Identification of Bacterial Isolates

The selected isolates were further identified by 16s rRNA gene analysis. Here >1000 nucleotide bases of 16s rRNA gene belonged to isolate LB 1.1 showcased best-scored homology with accession number LC 183868 belongs to *Lysinibacillus macroides*, NA + LB found to be better homolog with *Acinetobacter pittii* (NR_117621.1) with accession no LC155825 and analysis of M9 + LB shows similarity with *Bacillus safensis* (NR041794) with accession no LC155823. Phylogram designed with five top scorer sequences was also highlighting the same output as shown in following (Fig. 4a–c).

7.5 Chromium Reduction at Different Cr Concentration and Time

In the percent reduction assay of chromium carried out by Diphenyl Carbazide method, results highlighted that, *Lysinibacillus macroides* maintained at 300 mg/L and 500 mg/L recorded the early reduction of chromium at 24 h with 67% and 68.6% value when compared with 10 and 21% reduction recorded at 100–700 mg/L (Fig. 5a). Upon 48 h once again 300 mg/L and 500 mg/L of chromium set able to reduce 75% and 76% of chromium while 100 mg/L recorded with only 52% and 700 mg/L with 35%. It is important to note that at 300 mg/L and 500 mg/L isolate. *L macroides* able to reduced >90% of chromium at 72 h only while it took 120 h to achieve the same when isolate grown at low Cr concentration (100 mg/L). And at high concentration (700 mg/L) this isolate can able to reduce only up to 65% of chromium. Several studies have been showed *L. sphaericus* ability of tolerance and accumulation of heavy metals such as Pb (II), Cr (VI), Cd (II) [58–60].

In case of *Acinetobacter pittii* at 100 mg/L concentration of chromium probably with more dilution factor, it has affected the interaction of the isolate with the content of chromium and resultant even though concentration was low only 80% of maximum reduction was recorded. In case of 300 mg/L it has been observed that 100% Cr reduction was observed in presence of *A. pittii* and which was 95% at 500 mg/L and lowered down again to 75% at 700 mg/L (Fig. 5b).

Figure 5c shows the chromium reduction ability of *Bacillus safensis* (M9 + LB). Significant difference observed in the reduction at different concentration within 24 h. The highest reduction was observed at 300 mg/L concentration of chromium, with the percentage 89%, 94.66%, 97% and 99% in 24, 48, 72 and 72 h respectively, followed by 500 mg/L concentration with the percentage 68.66%, 76.66%, 82%, 91%, 95% in 24, 48, 72, 96 and 120 h respectively. The amount of Cr (VI) reduced at 100 mg/L was found to be 6%, 12%, 31%, 68% and 100% after 24, 48, 72, 96 and 120 h. The least reduction rate was observed at the highest concentration of Cr (VI) 700 mg/L with the percentage 18%, 43%, 53.33%, 70%, 72% in 24, 48, 72, 96 and 120 h respectively.

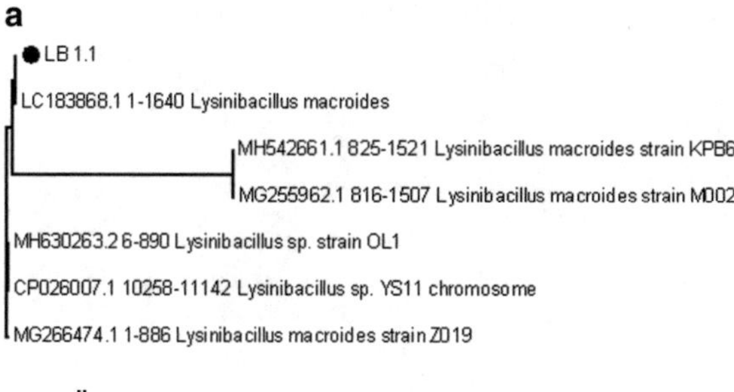

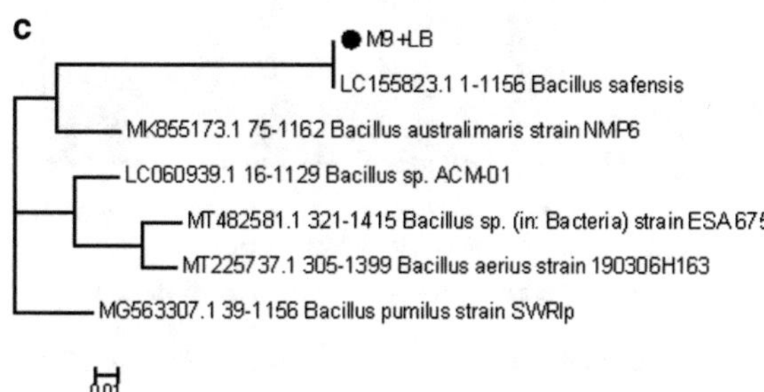

Fig. 4 **a** Phylogenetic Tree of *L. macroides* (LB 1.1), **b** Phylogenetic tree of *Acinetobacter pittii* (NA + LB), **c** Phylogenetic tree of *Bacillus safensis* (M9 + LB)

An interesting fact observed that, bacteria can tolerate up to 1100 mg/L but can reduce only 700 mg/L conc. of chromium. Thus, it could be said that Cr (VI) reduction capability of these isolates was not related to Cr (VI) tolerance ability.

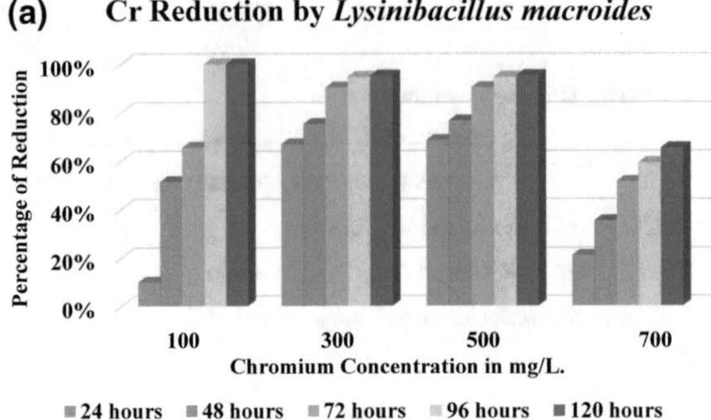

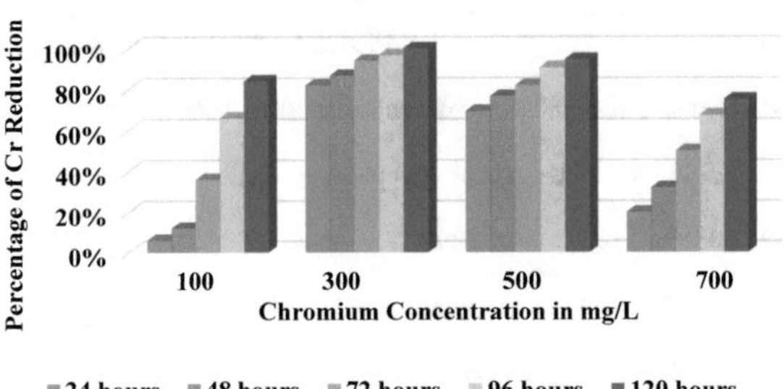

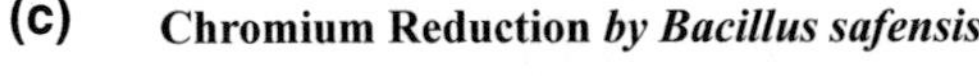

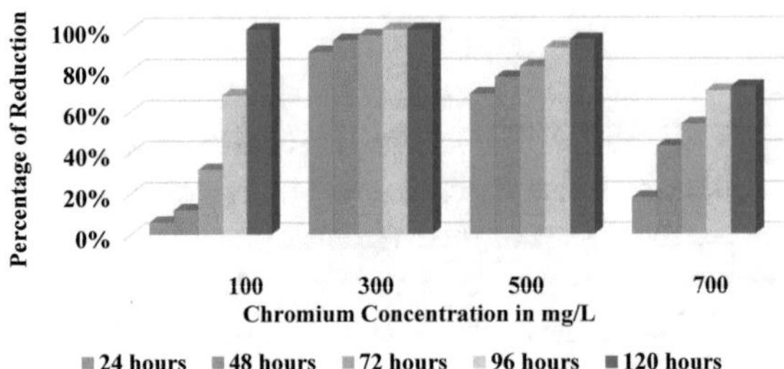

Fig. 5 **a** Reduction in chromium concentration by *Lysinibacillus macroides,* **b** Reduction in chromium concentration by *Acinetobacter pittii*, **c** Reduction in chromium concentration by *Bacillus safensis*

7.6 SEM-EDX Analysis

All three isolates when treated with chromium at variable concentration it has observed that isolate able to reduce chromium in the medium up to 100% at 120 h. These findings confirmed with SEM-EDX analysis. When these isolates comes in contact with chromium it can absorb chromium on their cell surface. This phenomenon is evident from SEM analysis where *L. macroides, A. pittii and B. safensis* shows distortion in their surface morphology with physiological alteration in the presence of chromium when compared to control as shown in Figs. 6, 8 and 10. In a parallel study, EDX analysis also confirmed that cell surface of all three isolates remains positive for chromium at a high level which was not at all evidenced in the control group as in Figs. 7, 9 and 11.

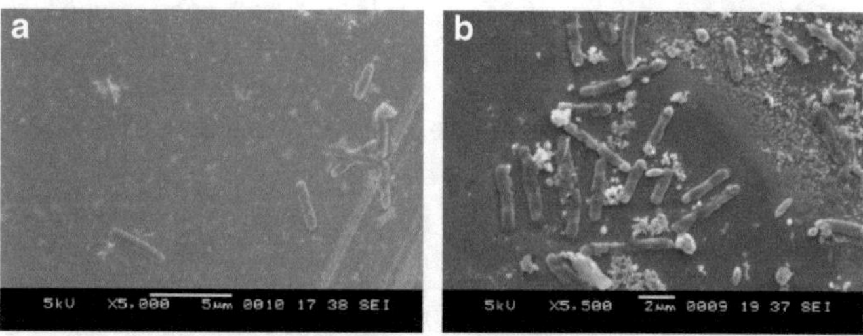

Fig. 6 **a** SEM Images of *Lysinibacillus macroides* with chromium **b** without chromium shows changes in the surface morphology of the bacterial cells

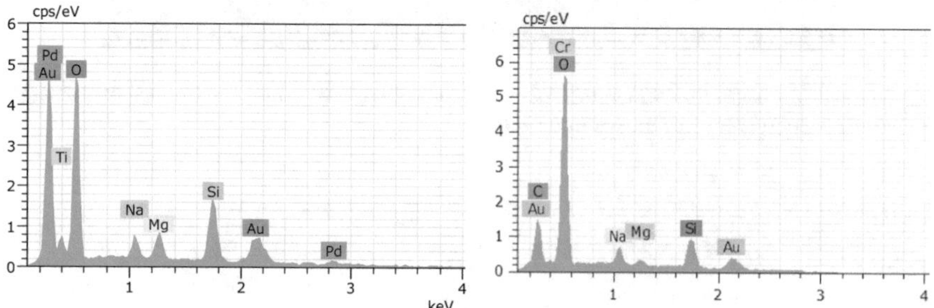

Fig. 7 EDX of *Lysinibacillus macroides* **a** in presence of chromium **b** in absence of Chromium

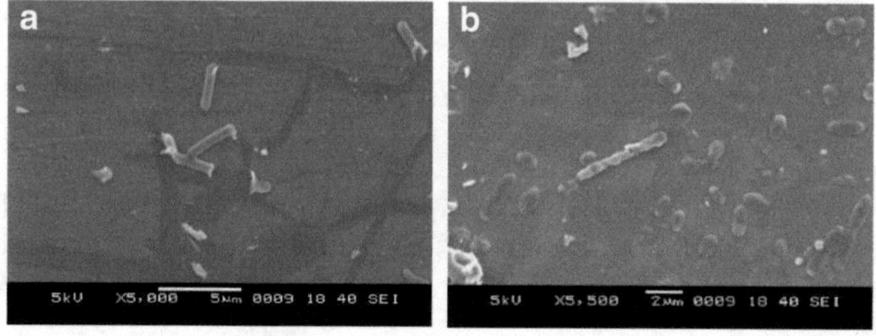

Fig. 8 **a** SEM Images of *Acinetobacter pittii* with chromium **b** without chromium shows changes in the surface morphology of the bacterial cells

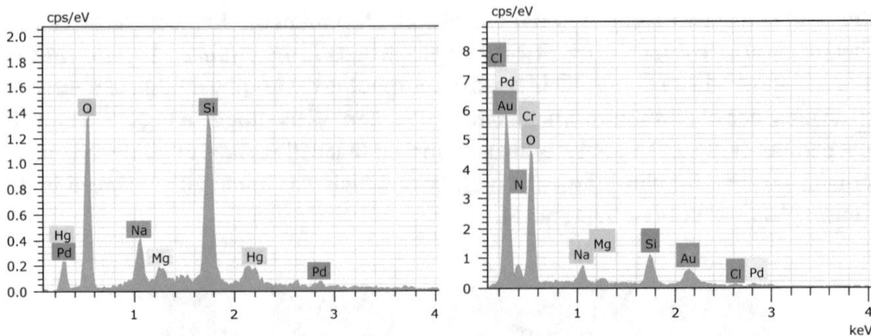

Fig. 9 EDX of *Acinetobacter pittii* **a** in presence of chromium **b** in absence of Chromium

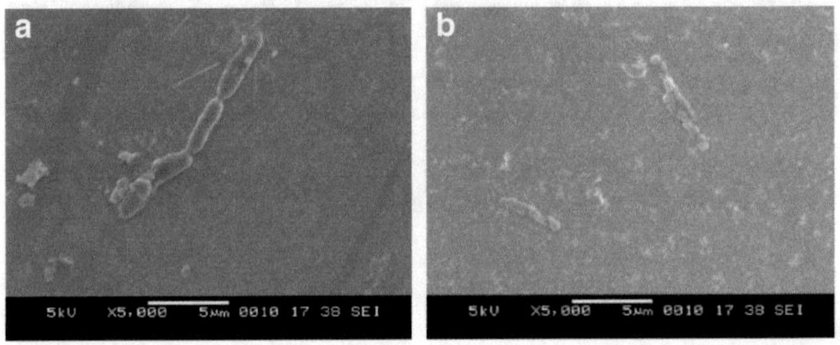

Fig. 10 **a** SEM images of *Bacillus safensis* with Chromium, **b** without Chromium shows changes in the surface morphology of the bacterial cells

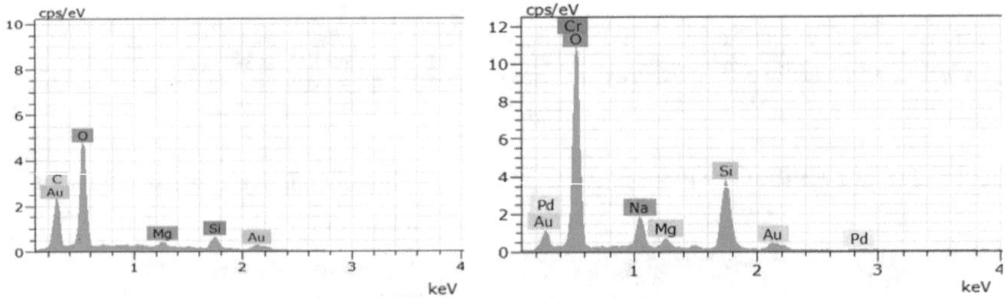

Fig. 11 **a** EDX in absence of Chromium, **b** EDS in presence of Chromium of *Bacillus safensis*

In a similar findings interaction of chromium adherence brings about changes in cellular morphology as well as on surface topology bacteria has been evidenced when investigated in *B. subtilis, Halomonas sp. Acinetobacter haemolytieus and sematia sp.* [61–64].

7.7 Chromium Reduction

L. macroides showcased multidrug-resistant features by remaining not-sensitive towards ampicillin, chloramphenicol, Co-trimoxazole, fusidic acid, linezolid, tobramycin and vancomycin. *A pittii* also showed resistance towards Ampicillin, Augmentin, Cefuroxime, Co-Trimoxazole, Fusidic Acid, Imipenem, Linezolid, Rifampicin, Tetracycline, and Vancomycin. Whereas *Bacillus safensis* showed resistance towards Augmentin and Vancomycin. It has been evident that the antibiotic resistance linked with ability of isolate to tolerate heavy metals as evidenced in study as well as early also [65]. Hence it is imperative to study antibiotic sensitivity profile along with metal tolerance study for bacteria spanning soil and related areas [58, 66].

In present study chromium mine of Pauni found to be positive for the *L. macroides, A. pittii and B. safensis* which showcase ability to grow in the presence on stressed condition of chromium up to 1100 mg/L. These strains were able to grow its population even though chromium added in the medium at the level of 100–1200 mg/L. These strains in return also showcased inherent antibiotic resistance even though it is that linked with the human pathogen list.

8 Conclusion

The present study successfully isolated culturable environmental bacterial species sampled from chromium mine soil of Pauni, Maharashtra. Isolates when identified by 16S rRNA gene sequencing showcase the best match with *Lysinibacillus macroides, Acinetobacter pittii and Bacillus safensis and* named as same. These isolates when tested in vitro, isolates retains the tolerance limit as high as 1100 mg/L and best reduction capacity at 300 mg/L minimum time and up to 500 mg/L at a maximum time with nearly 100% reduction of chromium.

Study put forward once again that by selecting proper sampling site for candidates isolation involved in metal bioremediation as in this case 'Pauni mine soil', chances of receiving promising isolates rises many folds. Study effectively identified the isolates found to be tolerating up to 900 and 1100 mg/L of chromium existence around them and also been able to reduce it up to 75% with concentration of 700, 300 mg/L was found to an optimum concentration for reduction in chromium concentration. Study reported these three isolates capable of chromium reduction from the nature by adsorbing it onto its cell surface and may be useful in coming time for effective bioaccumulation.

L. macroides and *A. Pittii* being a metal tolerant species also extend its ability towards antibiotic resistance which makes it a robust organism to use directly especially where human exposure occur directly or indirectly.

Lastly, all three isolates recorded to accumulate chromium on the cell surface as well as inside of it which makes this bacterium important in bioremediation approach and may find its utility in environment cleaning especially when chromium pollution occurs.

This type of study first time conducted in the chromite mine area of Nagpur and Bhandara to study isolation and identification of suspected emerging pathogen having heavy metal tolerance and multiple drug resistance.

It is a matter of concern as metal tolerance is also associated with antibiotic resistance. There is a need of regular health check-up and use of rational and proper antibiotic to avoid clinical outcomes. As these species have potential to reduce chromium it can be beneficial in the field of bioremediation.

Acknowledgements We would like to thank Department of Microbiology, Seth Kesarimal Porwal College, Kamptee, Nagpur, Maharashtra. We are also thankful to Department of Metallurgy, VNIT, Nagpur and IBM, Mineral Testing Lab, Hingana, Nagpur for their valuable support.

References

1. Koleli N, Demir A (2016) Chromite environmental materials and waste
2. Lunk JH (2015) Discovery, properties and applications of chromium and its compounds. Chem Texts 1:6
3. Pechova A, Pavlata L (2007) Chromium as an essential nutrient: a review. Veter Med 52:1–18
4. Spears JW, Whisnant CS, Huntington GB et al (2012) Chromium propionate enhances insulin sensitivity in growing cattle. J Dairy Sci 95:2037–2045
5. Mamyrbaev AA, Dzharkenov TA, Imangazina ZA, Satybaldieva UA (2015) Mutagenic and carcinogenic actions of chromium and its compounds. Environ Health Prev Med 20(3):159–167
6. Wilbur S, Abadin H, Fay M, Yu D et al (2012) Toxicological Profile for Chromium. Atlanta (GA): Agency for Toxic Substances and Disease Registry (US)
7. Lim JH, Kim HS, Park YM et al (2010) A case of chromium contact dermatitis due to exposure from a golf glove. Annal Dermatol 22(1):63–65
8. Bregnbak B, Jeanne D, Morten S, Zachariae C et al (2015) Chromium allergy and dermatitis: prevalence and main findings. Contact Dermat 73:261–280
9. Buters J, Biedermann T (2017) Chromium (VI) contact dermatitis: getting closer to understanding the underlying mechanisms of toxicity and sensitization. J Invest Dermatol 137:274–277
10. Costa M, Hong S, Brocato J (2015) Oral chromium exposure and toxicity. Curr Envir Health Rpt 2:295–303
11. Indian Minerals Yearbook (2017) (Part- III: Mineral Reviews) 56th Edition chromite. Government of india, ministry of mines, indian bureau of mines
12. Lunk H-J (2015) Discovery, properties and applications of chromium and its compounds. Chem Texts 1:1–17
13. Sahu KC, Godgul G (1995) Chromium contamination from chromite mine. Environ Geol 25:251–257
14. Tiwary RK, Dhakate R, Ananda V, Singh VS (2005) Assessment and prediction of contaminant migration in ground water from chromite waste dump. Environ Geol 48:420–429
15. Shanker K, Cervantes C, Tavera H, Avudainayagam S (2005) Chromium toxicity in plants. Environ Int 31:739–753
16. Rai V, Vajpayee P, Singh S, Mehrotra S (2004) Effect of chromium accumulation on photosynthetic pigments, oxidative stress defense system, nitrate reduction, proline, level and eugenol content of Ocimum tenuiflorum L. Plant Sci 1159–1169
17. Shahid M, Shamshad S, Rafiq M, Khalid S et al (2017) Chromium speciation, bioavailability, uptake, toxicity and detoxification in soil-plant system: a review. Chemosphere 178:513–533

18. Smith S, Peterson P, Kwan K (1989) Chromium accumulation, transport and toxicity in plants. Toxicol Environ Chem 24:241–251
19. Panda SK, Choudhury S (2005) Chromium stress in plants Braz. J Plant Physiol 17:95–102
20. Aktaruzzaman M, Fakhruddin A, Chowdhury M et al (2013) Accumulation of heavy metals in soil and their transfer to leafy vegetables in the region of dhaka aricha highway, savar, bangladesh Pakistan. J Biol Sci 16:332–338
21. Datta J, Bandhyopadhyay A, Banerjee A (2011) Phytotoxic effect of chromium on the germination, seedling, growth of some wheat (Triticum aestivum L.) cultivars under laboratory condition. J Agricul Technol 7:395–402
22. Nagarajan M, Sankar K (2014) Effect of chromium on growth, biochemicals and nutrient accumulation of paddy (Oryza sativa L.). Int Lett Natural Sci 23:63–71
23. Andaleeb F, Anjum M, Ashraf M, Khalid Z (2008) Effect of chromium on growth attributes in sunflower (Helianthus annuus L.). J Environ Sci 20:1475–1480
24. Vajpayee P, Tripathi RD, Rai U, Ali M, Singh S (2000) Chromium (VI) accumulation reduces chlorophyll biosynthesis, nitrate reductase activity and protein content in Nymphaea alba L. Chemosphere 41:1075–1082
25. Hooda S, Yibing S (2010) Chromium, Nickel and cobalt: trace elements in soil. s.l. Wiley, pp 19–471
26. Zhang X, Zhang X, ZhangL Chen Q, Yang Z et al (2012) XRCC1 Arg399Gln was associated with repair capacity for DNA damage induced by occupational chromium exposure. BMC Res Notes 5:1–7
27. Gibb HJ, Lees PS, Pinsky PF, Rooney BC (2000) Clinical findings of irritation among chromium chemical production workers. Am J Ind Med 38(2):127–131
28. Halasova E, Matakova T, Kavcova E et al (2009) Human lung cancer and hexavalent chromium exposure. Human lung cancer and hexavalent chromium exposure. Neuro Endocrinol Lett 182–185
29. Singh VP (2005) metal toxicity and tolerance in plants and animals. SARUP & SONS, New Delhi
30. Pan X, Hu J, Xia W, Zhang B et al (2017) Prenatal chromium exposure and risk of preterm birth: a cohort study in Hubei, China. Nature Sceint Rep 7:1–8
31. Saxena DK, Murthy RC, Lal B et al (1990) Effect of hexavalent chromium on testicular maturation in the Rat. Reprod Toxicol 4:223–228
32. Assem L, Zhu H (2007) Chromium -Toxicological overview. Institute of Environment and Health, Cranfield University, pp 1–14
33. Igiri EB, Okoduwa SR, Idoko GO et al (2018) Toxicity and bioremediation of heavy metals contaminated ecosystem from tannery wastewater: a review. J Toxicol
34. Bruins MR, Kapil S, Oehme FW (2000) Microbial resistance to metals in the environment. Ecot Environ Saf 198–207
35. Zahoor A, Rehman A (2009) Isolation of Cr (VI) reducing bacteria from industrial effluents and their potential use in bioremediation of chromium containing wastewater. J Environ Sci 21:814–820
36. Congeevarama S, Dhanarani S, Park J et al (2007) Biosorption of chromium and nickel by heavy metal resistant fungal and bacterial isolates. J Hazard Mater 146:270–277
37. Acevedo-Aguilar FJ, Espino-Saldaña AE, Leon-Rodriguez IL et al (2006) Hexavalent chromium removal in vitro and from industrial wastes, using chromate-resistant strains of filamentous fungi indigenous to contaminated wastes. Canadian J Microbiol 52:809–815
38. Baldi F, Vaughan AM, Olson GJ (1990) Chromium (VI) resistant yeast isolated from a sewage treatment plant receiving tannery wastes. Appl Environ Microbiol 56:913–918
39. Han X, Shan YW, Wong MH, Tama NFY (2007) Biosorption and bioreduction of Cr (VI) by a microalgal isolate, Chlorella miniata. J Hazard Mater 146:65–72
40. Dixit RW, Malaviya D (2015) Bioremediation of heavy metals from soil and aquatic environment: an overview of principles and criteria of fundamental processes 7:2189–2212

41. Wu G, Kang H, Zhang X et al (2010) A critical review on the bio-removal of hazardous heavy metals from contaminated soils: issues, progress, eco-environmental concerns and opportunities. J Hazard Mater 174:1–3
42. Schulz S, Brankatschk R, Dumig A et al (2013) The role of microorganisms at different stages of ecosystem development for soil formation. Biogeosciences 10:3983–3996
43. Konovalova VV, Dmytrenko GM, Nigmatullin RR et al (2003) Chromium (VI) Reduction in Membrane Bioreactor with Immobilized Psedo-monas Cells. Enzyme Microbial Technol 33:899–907
44. Garbisu C, Alkorta I, Lama MJ, Serra JL (1998) Aero-bic Chromate Reduction by Bacillus Subtilis. Biodegra-dation 9:133–141
45. Ishibashi Y, Cervantes C, Silver S (1990) Chromium reduction in Pseudomonas putida. Appl Environ Microbiol 56:2268–2270
46. Michel C, Brugna M, Aubert C (2001) Enzymatic reduction of chromate: comparative studies using sulfate-reducing bacteria. Appl Microbiol Biotechnol 55:95–100
47. Wang P, Mori T, Toda K, Ohtake H (1990) Membrane associated chromate reductase activity from Enterobacter cloacae. J Bacteriol 172(3):1670–1672
48. Wang H, Shen YT (1993) Characterization of enzymatic reduction of hexavalent chromium by eshcherichia coli ATCC 33456. Appl Environ Microbiol 59:3771–3777
49. Chaturvedi MK (2011) Studies on chromate removal by chromium-resistant bacillus sp. isolated from tannery effluent. J Environ Protect 76–82
50. Batool R, Yrjala K, Hasnain S (2012) Hexavalent chromium reduction by bacteria from tannery effluent. J Microbiol Biotechnol 22(4):547–554
51. Kipkurui NL, Salim AM, Nyambati V et al (2016) Isolation and molecular characterization of chrome resistant bacteria from chrome contaminated tannery waste from disposal sites in kenya. IOSR J Appl Chem 9:1–5
52. Clausen CA (2000) Isolating metal tolerant bacteria capable of removing copper, chromium, and arsenic from treated wood. Waste Manag Res 264–268
53. Environmental Investigations Standard Operating Procedures and Quality Assurance Manual (2001) U.S. Environmental Protection Agency
54. Bauer AW, Kirby WMM, Sherris JC, TurckM (1966) Antibiotic susceptibility testing by a standardized single disc method. Am J Clinical Pathol 45:493–591
55. Oyetibo GO, Ilori MO, Adebusoye SA et al (2012) Bacteria with dual resistance to elevated concentrations of heavy metals and antibiotics in Nigerian contaminated systems 168(1–4):305–314
56. Hudzicki (2009) Kirby-bauer disk diffusion susceptibility test protocol. Am Soc Microbiol
57. Mupidwar NA, Ingle AB, Magar SP (2015) Prevalence of antimicrobial resistant E. coli in water reservoirs. J Pharm Res 9(8):522–524
58. Velásquez L, Dussán J (2009) Biosorption and bioaccumulation of heavy metals on dead and living biomass of Bacillus sphaericus. J Hazard Mater 167(1–3):713–716
59. Montenegro P, Dussán J (2013) Genome sequence and description of the heavy metal tolerant bacterium Lysinibacillus sphaericus strain OT4b.31. Stand Genom Sci 9:42–56
60. Pal A, Datta S, Paul A (2013) Hexavalent chromium reduction by immobilized cells of bacillus sphaericus AND 303. Brazilian Archives Biol Technol 56:502–515
61. Mangaiyarkarasi MS, Vincent S, Janarthanan S et al (2011) Bioreduction of Cr (VI) by alkaliphilic Bacillus subtilis and interaction of the membrane groups Saudi. J Biol Sci 18:157–167
62. Mabrouka M, Arayesa M, Sabry S (2014) Hexavalent chromium reduction by chromate-resistant haloalkaliphilic Halomonas sp. M-Cr newly isolated from tannery effluent. Biotechnol Biotechnol Equip 28:659–667
63. Zhang K, Li F (2011) Isolation and characterization of a chromium-resistant bacterium Serratia sp. Cr-10 from a chromate-contaminated site. Appl Microbiol Biotechnol 90:1163–1169
64. Pei QH, Shahir S, Raj ASS et al (2009) Chromium (VI) resistance and removal by Acinetobacter haemolyticus World. J Microbiol Biotechnol 25:1085–1093

65. Knapp CW, Callan AC, Aitken B (2017) Relationship between antibiotic resistance genes and metals in residential soil samples from Western Australia. Environ Sci Pollut Res 24:2484–2494
66. Austin CB, Wright M, Stepanauskas R, McArthur JV (2006) Co-selection of antibiotic and metal resistance. Trends Microbiol 14:176–181

Managing Water Quality in Mining Areas: Changing Paradigm of Sustainability

Rajani Tumane, Shubhangi Pingle, Aruna Jawade, and Kirtikumar Randive

Abstract Water is an important resource and a necessity for the mining and minerals processing industries. The mining and mineral-based industries extract water from groundwater, streams, rivers, and lakes, for mineral processing, dust suppression, and slurry transport. When the available water in mines is in excess, it can cause flooding in the underground mines; and when it is scarce, it can significantly influence the mining operations. Wastewater generated by the underground and open-cast mines and acid mine drainage produced from the oxidation of metal sulfides, contains higher levels of dissolved aluminium (Al), iron (Fe), and sulfate (SO_4). The oxidative dissolution of sulfide minerals releases extremely acidic leachates consisting of toxic elements such as sulfur, arsenic, cadmium, copper, chromium, nickel, lead, uranium, and zinc, which can potentially affect the surrounding environment. The interactions of the elements associated with acid mine drainage is necessary to study for developing proper wastewater management strategy. Studies have shown that the mining activity influences the groundwater and also contaminates water bodies in proximity. In India, water discharge standards/limits containing 33 parameters were framed under the Environment (Protection) Rule, 1986 (under Schedule VI) by the Ministry of Environment, Forest and Climate Change (MOEFCC). Mining wastewater management from surface, underground, and acid mine drainage is therefore one of the most challenging factors in the mining sector. There were several techniques available to treat physiochemical properties of polluted water such as screening, skimming, sedimentation, neutralization, equalisation, oxidation pond,

---

R. Tumane (✉)
ICMR–National Institute of Occupational Health, Ahmedabad, Gujarat, India
e-mail: rajo_tumane@yahoo.co.in

S. Pingle · A. Jawade
ICMR–Regional Occupational Health Centre (Southern), Bangalore, (KA), India
e-mail: pingle.shubhangi@gmail.com

A. Jawade
e-mail: ajwade28@rediffmail.com

K. Randive
Department of Geology, Rashtrasant Tukadoji Maharaj Nagpur University, Nagpur 440001, (MH), India
e-mail: randive101@yahoo.co.in

K. Randive et al. (eds.), *Innovations in Sustainable Mining*, Earth and Environmental Sciences Library, https://doi.org/10.1007/978-3-030-73796-2_12

activated sludge process, trickling filters, chlorination, and reverse osmosis. The Total Dissolved Solids (TDS), Total Suspended Solid (TSS), Biological Oxygen Demand (BOD), Dissolved Oxygen (DO), Chemical Oxygen Demand (COD), are the most common parameters that are monitored and treated by the mining companies. In order to meet the present as well as future demands for creating the effective water management strategies, the mining companies need to be increasingly conscious of the risks and opportunities involved in the management of water resources. Although the modern water management systems have better modelling tools, these systems need improvement in their online monitoring tools for data collection, organization, and transfer. To secure quantity as well as quality of water throughout the life cycle of a mine, it is important to make optimum utilization of available resource and ensure its recycling. This chapter provides a critical review of the issues in the management of mine waters and their influence on the health and environment.

Keywords Acid mine drainage · Ground water · Metal toxicity · Surface water · Contaminants · Recycling processes

1 Introduction

The social demand for minerals has risen in general due to rapid industrialization and socio-economic growth of the nation [1]. This mad-rush for exploitation of mineral resources left unprecedented stress on the environment, most of which is irreversible in the foreseeable future [1–4]. The Indian mining industries which produced 89 minerals under different categories which consists of 4 fuels, 11 metallic, and 52 non-metallic industrial and 22 minor minerals [6]. In the world, India ranked 2nd in barite, chromite, and talc, 3rd in coal and lignite, 4th in iron ore and kyanite, 6th in bauxite, and 7th in manganese production [2, 4]. In India, 80% of mining is coal and 20% in various metals and other raw materials including gold, copper, iron, lead, bauxite, zinc and uranium. Minerals occur in India as metallic forms which consist of the ferrous group (Iron, Manganese, Chromium); non-ferrous (Aluminium, Copper, Lead, Zinc) [5–7].

The metallurgical and coal mining requires great deal of water for different mining activities e.g., coal washing, releases toxic metals with mine wastewater in the surrounding and river etc. [8]. The opencast and underground mining processes, generated huge amount of solid wastes, tailings, slurry, mine water, rock debris, slag, vegetation, and overburden of mine having direct directly influence on the hydrology and ecosystem [9]. Acid mine drainage (AMD) occurs around coal and other metallic mines having sufficient quantities of sulphide minerals and sulphide contents. Due to AMD, the surface and groundwater gets contaminated resulting into disruption of growth and reproduction of aquatic plants and animals occurring around mining areas [10].

Studies observed that mining activity deteriorates water quality and quantity in nearby areas of industrial establishments depending on the hydrological, geological, and environmental conditions of the area [11]. The flora and fauna of the surrounding ecosystems have been severely affected due to hazardous toxicants and metals [12]. Presently, untreated mining drainage, crippling effects of fluoride and arsenic toxicity, paucity of clean and safe water for drinking and farming have become a herculean task for authorities [13]. These contaminants must be tactically treated and managed to combat economic efficiency and demand for environmental sustainability [14].

World over, the water legislation has been effective and aptly supported by the administrative machinery. A variety of useful designing and monitoring tools have been recently introduced, which effectively provide a balance between the growing water demands and the need to safeguard and secure groundwater resources for future generations [15]. Nevertheless, the challenge of water security for mining and mineral processing industries is critical. Although new technologies and techniques are available for environmental rehabilitation, treatment of mine drainage, and management; most of these technologies have not been implemented in India due to variety of problems including, differences in geography, climate, finance, and management.

On this backdrop, present chapter is aimed at shedding some light on the requirements, quantity, quality and management of mine wastewater with special emphasis on the sustainability and application of mine waste water in the vicinity of mines.

2 Ground and Surface Water Pollution

Opencast mining creates huge quantities of gangue i.e., the waste rock material. Mining activity disrupts original water flow channels that were in balance before the excavation. The natural springs that were live connections between the surface and groundwater are vanished, either due to impoverishment or excessive drawdown of water during mining activities. Thus, the overall water balance of an area is severely affected [16]. Wastewater from an open-pit mine is discharged, intercepting the water table, and the water from the phreatic surface (water table) is siphoned out creating a 'cone of depression' [17]. The cone of depression enlarges due to continuous removal of water from a particular area (e.g., a dug-well, bore-well, or a mining pit), if the water is discharged for longer time. This is known as the 'drawdown'. This has a long-lasting impact on the environment and ecology of the area [18–20].

National Environment Engineering Research Institute (2002) report has shown that the fine dust particles from the dumping site of mines near the Bellary-Hospetmine get dispersed along the hill slopes through surface runoff and enters into nearby water bodies during monsoon. Similarly, the waste rock dumps and the freshly cut bedrock walls were also the principal sources of metal pollution due to mining [21]. High levels of arsenic, sulfuric acid, and mercury were reported in surface or groundwater [22]. A recent study conducted in the mines operated by Coal India Limited has indicated that the pollutants exceed the limits prescribed by the Bureau of Indian Standards in 8 out of the 28 mines [23]. The surface water includes streams, lakes, ponds, and ocean waters were degraded by disposal or spillage of toxic chemicals, eroded wastes, and discharge of mine water. In India, about 70% of groundwater and surface water were jeopardized in quality due to mining and other industrial activities [24]. The mine water has low pH (acidic) and high amount of turbidity, iron, sulfate, and total hardness. This water is capable of dissolving metals, such as copper, aluminum, cadmium, arsenic, lead, and mercury in the active mining areas. Similarly, it also leaches these metalliferous constituents from the mineralized rocks. At an appropriate Eh-pH conditions, the metal-saturated waters get precipitated forming duricrusts on the stream bottom. This iron-rich, orange-red colour slime is called as 'yellow boy' [25]. The mine waters are neither potable not suitable for agriculture due to their higher concentration of iron and manganese. Moreover, the mining operations such as drilling and pumping can mobilize arsenic or fluoride, induce saline intrusion, or influence the migration of lower quality or brackish waters [26]. The coal washeries spill-out thousands of litters of industrial chemicals into the rivers, thereby jeopardizing the drinking water quality of the surrounding areas. On the other hand, the surface water quality can be severely affected by the erosion of solid wastes, tailing ponds, open pits, ore heap, dump leaches, overburden piles, haulage roads and access roads, ore stockpiles, vehicle, and equipment maintenance areas,, and reclamation sites [24, 27]. If the solid wastes are not properly stored, the toxic metals can be drained by the streams into water reservoirs, where they react with oxygen and water to release toxic metals and deleterious constituents into the streams [28, 29]. Hydraulic connection between groundwater and surface water quality was affected by various factors such as surface hydrology, evaporation, soil texture, and terrestrial vegetation when wastewater infiltrates through surface water into groundwater. Dissolved metals in surface water can cause a lesion in the gills of fish and also affect the aquarium life of microorganisms [29, 30].

3 Acid Mine Drainage (AMD)

Mining activities create a disturbance on the surface level, drainage system, mining operation, acid mine drainage, and withdrawing groundwater for mine working. Acid mine drainage (AMD) is the most difficult mine waste problem generated at both abandoned and active mine sites. AMD gets mixed with neighbouring watercourses and disrupts the physical, chemical, and biological characteristics of the natural water resources [31]. The AMD occurs as a result of oxidation of sulfides, especially pyrite, at low pH and high levels of SO_4^-, iron, metalloids, and other metals. The natural drainage systems also get affected by AMD due to physico-chemical equilibrium and precipitation of As, Cd, Cu, Ni, and Pb meters downstream from the acid discharge, removing chemical species [32]. Recently, studies were carried out to understand the physicochemical equilibrium of mining wastewater and Acid mine drainage (AMD) forms in sulfide minerals or other sulfide minerals such as sphalerites (ZnS(S)), covellite (CuS(S)), millerites (NiS(S)), galena (PbS(S)), greenockite (CdS(S)) and white pyrites(FeS_2) were directly exposed to oxidizing conditions in metal mining [33]. According to Jennings et al. [34], precipitation of 'yellow boy" Fe $(OH)_3$ changes the color and turbidity of natural water resources, fisheries life cycle, benthic algae, invertebrates, fish, and photosynthesis of green plants because of discharge of mining and industrial wastewater in the river, lake pond [34].

4 Treatment of Acid Mine Drainage by Active and Passive System

Treatment of mining wastewater or AMD can be done by the active and passive systems. The active treatment can be carried out in four stages in fixed plants. An active treatment system required less construction area as compared to a passive treatment system such as a constructed wetland). These systems may remove metals, cyanide components, solid-liquid separation, and sludge removal through coagulation, gravity separation, and filtration processes. Finally, treated water is discharged into an effluent water pond for further use in the mining system. An inactive system, failure of equipment, and budget can result in lapses in the biological and physical degradation of the streams [35]. A man-made passive treatment system can treat the effluent without constant human intervention. An artificial wetland in which organic matter, such as hydroxide and sulfide precipitation, or pH adjustment, bacteria, and algae work together to remove the heavy metal ions and decrease the soil redox potential by the process of filtrations, adsorptions, and precipitation. Finally, water was passed through a series of interconnected ponds. Therefore, their use is preferable during the mine's post-closing stage or for low acidity AMD. To choose the proper passive treatment method, AMD characteristics need to be quantified. This happens by quantifying the 'Acidity' and 'Acidity Load' which was linked with the concentration of calcium carbonate ($CaCO_3$) since the dissolution of this substance in water affects the pH change causing acidity [36–41].

There are risks associated with the processing and treatment of waste waters, which are broadly categorized into following four: (i) treatment risks, (ii) effluent management risks, (iii) sludge management risks, and (iv) risks due to natural disasters. In the first category, the risks due to mechanical failure, power failure, plugging of the substrate, piping or ditches, armoring of reactants, failure of reagent delivery system, failure of process control components, improper design volume of holding ponds, scaling of plant components, and shut down due to labour disruption are included. In the second category risks such as dissolved metals, pH, effluent toxicity test failure, change in permit requirement, and inability to meet required environment water quality standard. In the third category, the risks due to sludge density, lack of appropriate on-site disposal, off-site transportation, and disposal issues, poor sludge stability (chemical mobilization, physical instability), sludge pond access risks (human/fauna), and dusting (airborne contamination) are included. In the fourth category the risks are predicted due to natural disasters such as earthquakes, tsunamis, excessive precipitation, hurricane [42–44]. Although, the above factors are called as risks, this may not be an appropriate word to describe all above factors. In any given operations, these are predictable. These are better be referred to be the 'challenges' than 'risks'; however, we leave it there for the individuals to use appropriate jargon.

Recently, mining-based industries used limestone, wetlands, filtration, and neutralization processes for the purification of wastewater. The design and geometry of the limestone ponds were constructed in such a way that the bottom portion of the pond and mine water flows upward through the limestone. It was also based on the topography of the area and the geometry of the discharge zone. This pond is suitable for low dissolved oxygen (DO) containing water and free of Fe^{3+} and Al^{3+} radicals. The wetlands (CWs) purification system, require fewer maintenance costs, robust operation, and fewer by-products generation. It's design and construction mainly depend upon the rate of loading, detention time, surface slope, substrate, types of vegetation, sediment control. Wetlands provide adsorption, neutralization, and precipitation that need to be incorporated into the hydrology, soil, and vegetation. The filtration process is a very effective, low-cost, and efficient pre-treatment process used for the removal of fine solid particles without the addition of any chemicals in mine water. This treatment process does not require skilled labor, operation, maintenance, effectiveness, and treatment efficiency [45–48]. Lime neutralization/precipitation are well-accepted processes for the purification of mine water in coal and metal industries. Mine water treatment management includes ammonia, caustic soda, calcium peroxide, hydrated lime, kiln dust, limestone, soda ash, and fly ash. In this treatment, mine wastewater is discharged into a rapidly mixed chamber where hydrated lime is added in dry or slurry form. These neutralization processes effectively reduced the concentration of ferrous iron at some suitable pH levels. This treatment depends on the physical, chemical, and biological characteristics of the mine water [3, 49].

5 Characteristic of Mine Wastewater and Chemical Properties of Polluted Water

Industrialization and mining are important pillars for growth and progression of countries. However, these activities come with a cost, such as change of topography, loss of biomass, excessive erosion and so on [50]. Quality of water has direct consequences on the quality of life and the people at large. It is very essential to ensure the quality of water before being used for drinking, domestic, agricultural, or industrial purposes [51, 52]. Water must be tested with different Physico-chemical parameters. The selection of parameters for testing of water, however, solely depends upon the purpose for which the water will be used and to what extent its quality and purity are needed. The identification of sources of wastewater and their characterization and remediation are crucial [53, 54]. The wastewater from a mine sites can be categorised into; mine water, processed wastewater, domestic wastewater, and surface run-off. Various deleterious constituents are known to occur in the mine waters and tailings, which include: coal, oils and grease, soaps and detergents, rubber, dyes, and phenolic compounds; Cr, Hg, Cu, Cd, Pb, Zn, Ni; acids, alkalis, cyanide; Mg, Ca, K, Na, Fe,

Table 1 Methods used for analysis of water quality [56]

S. no.	Test	Analysis method	Standards WHO limits for drinking water [61]
1	pH	Potentiometeric	6.5–9.2
2	Total alkalinity	Titrimetric to pH = 4.5 (methyl orange)	200
3	Boron	Curcumin spectrophotometric	1.0
4	Calcium	EDTA titrimetric	200
5	Chloride	Argentometric Titration	600
6	Colour	Visual comparison	25
7	Fluoride	Ion selective electrode	1.0–1.5
8	Total hardness	EDTA titrimetric	–
9	Iron	Phenanthroline Spectrophotometric	1.0
10	Manganese	Persulphate spectrophotometric	0.5
11	Nitrate	Selective electrode	100
12	Turbidity	Nephelometric	25
13	Dissolved solids	Gravimetric after filtration	500

Mn; Cl, SO_4^{2-}, NO_3^-, HCO_3^-, Pb_4; biological bacteria, viruses, and small organisms, radiological uranium, tritium, and other radioactive substances [14, 55–60]. Table 1 summarizes common parameters used for assessing the water quality, their common methods of analysis, and permissible limits for drinking.

6 Technologies to Combat Water Quality in Mining

Established new standard technologies typically followed a development process that leads from laboratory and bench-scale investigations to pilot studies and to initiate use or "full-scale demonstrations" before the technology was considered. These technological innovations are necessary for meeting the challenge of progressive wastewater pollution abatement, which is likely to increase due to population growth, changes in industrial processes, and technological developments. To understand the inherent problems of groundwater caused due to mining activity, the geohydrological setup of the area has to be studied first. The growing health concern has been an increasing awareness about the need to dispose of this wastewater safely and beneficially. Ever since the concerns of water pollutions came to the fore the mechanisms of wastewater treatment revolved around physical, chemical, and biological processes, in order to remove the solids, organic matter, and nutrients from wastewater. The wastewater treatment is categorized into preliminary, primary, secondary, and advanced methods [62]. Preliminary treatment includes removal of coarse solids and other big suspended materials found in raw wastewater.

Preliminary treatment process includes removal of solids and big size material occurred in raw wastewater. These techniques include (a) Screening (Suspended matter, floating objects including cloth, wood, plastic bags, etc. are allowed to pass through a screen made of iron bars with a spacing of 1 to 2 inches, (b) Skimming oil, grease, and similar impurities on the surface of the water are removed by skimming. For this, the compressed air is sent through a diffuser plate located at the bottom of the skimming tank).

6.1 Primary Treatment

Removal of Organic and Inorganic solids by sedimentation and floating (Scum) by skimming. 25–50% BOD, 50–70% suspended solids and 65% of the oil and grease is removed by this technique. Organic nitrogen, phosphorus, and heavy metal were also removed by this technique. Following methods are used: (a) Sedimentation (colloidal, solid and sludge, large objects from influent sewage can be removed by sedimentation. In this method, a sedimentation tank is constructed which is shallow and radial in shape. The depth of the tank is about 3–5 m and the diameter is ~30 m. The polluted water is kept in the tank for 3 to 4 h, and then accelerated by mechanical flocculation and chemical coagulation), (b) Mechanical flocculation (the wastewater is passed through a tank having a retention time of 30 min and rotating speed of 0.5 m/s with paddles. By this gentle stirring, the divided suspended solids come together to form larger particles, and then settle down easily), (c) Chemical Coagulation (this technique is used for the removal of suspended and colloidal particles present in water. For this purpose, hydrated lime, alum, ferric chloride, and a mixture of ferric sulfide and chloride are used), (d) Neutralisation (highly acidic or alkaline water is neutralized by treatment with limestone or caustic soda and H_2SO_4. (e) Equalisation (wastewater coming from the tank is mixed thoroughly and a homogeneous mixture is formed after which, the mixture is treated by sedimentation).

6.2 Secondary Treatment

In this process, the colloidal as well as organic matter from wastewater is removed. It removes ~70–80% of BOD by biological treatment using bacteria and other micro-organisms. Common processes involved include: (a) Oxidation Pond or stabilization pond (This is the simplest system used for treatment of wastewater, which uses algae present in the system [62]. A small is constructed with embarkments from all the sides to prevent entry of rainwater inside the pond), (b) Activated Sludge Process (the polluted water from the sedimentation tank is sent to the aeration tank, wherein, ~20–30% of the active sludge is thoroughly mixed for 4–8 h. This water to carried to secondary sedimentation tank for degradation and decomposition of organic matter by the microorganisms), (c) Trickling Filters (sprinkling filter) orbiofilter (these filters are cylindrical structures of 2 m depth and filled with the granular materials such as clinker, stone, coal gravels. The wastewater is sprinkled through this media and the organic matter present in the wastewater is metabolized by the biomass (d) Rotating Biological Contractor (RBCs) (The RBCs are rotating biological filters, consisting of a series of closely spaced plastic circular disks, which are slowly rotated. The micro-biotic substances in the wastewater are consumed by the biomass grown on the rotting surface of the discs, thereby decreasing their concentration from the wastewater).

6.3 *Tertiary Treatment/Advanced Wastewater Treatment*

This technique removes suspended solids, phosphorus, nitrogen, and inorganic impurities at the molecular level which was not removed at primary and secondary treatment. The mining effluent from the primary clarifiers flows to the biological reactor which is physically divided into five zones by baffles and weirs. Phosphorus removal process, anaerobic fermentation zone used the group of bacteria by stressing them under low oxidation-reduction conditions. Finally, released of phosphorus equilibrium in the cells of the bacteria with an adequate supply of oxygen and phosphorus in the aerated zones. Cells get quickly build up phosphorus in excess of their normal metabolic requirements and removed phosphorus from the system with the waste activated sludge. The nitrogen (ammonia) in the influent was removed through the first two zones. First, in anaerobic zone, the sludge is treated with ammonia nitrogen and converted into nitrite to nitrate. The nitrate-rich mixed liquor is recycled with aerobic zone back to the anoxic zone. Denitrification process involves recycling of nitrates with no oxygen and reduction by facultative bacteria to nitrogen gas and nitrogen gas escaped into the environment. In the anoxic zone, nitrates that are not recycled are reduced by the endogenous respiration of bacteria. Damage settling in the secondary clarifiers is mixed with liquor and flows. Oxygen levels are again raised to prevent further de-nitrification in the re-aeration zone [63]. Following processes are used for treatment: (a) Microstraining (removes solid suspended particles in the last advanced treatment plant that were not removed during the primary and secondary treatments. A rotating drum type filter is used to screen the suspended solids using the finely woven stainless-steel fabric with a mesh size of 23–25 μ mounted on the periphery and water is allowed to pass from inside out), (b) Chlorination (The hypochlorous acid is an effective purifying agent, which disinfects water from pathogenic bacteria and controls the growth of undesirable algae in the water. In this, the gaseous chlorine, bleaching powder, or sodium hypochlorite (NaOCl) is used to generate the desired compound), (b) Coagulation sedimentation (For removal of phosphorous from wastewater, alum, lime or iron salts are added. With the addition of these chemicals, the small size clump and 'floc' join together to form into big size masses. The big size masses are allowed to settle out in the sedimentation tank and reduce the concentration of phosphorus by about 95%), (c) Adsorption (It is a physical process for removing low concentrations of contaminants from water. Activated carbon has been processed to create a very large surface area as great as 1500 m^2/g (7.3 million ft^2/lb) available for the adsorption of contaminants. The porous nature of adsorbent makes available large surface area and also provides cavities for blocking of contaminants), (e) Reverse Osmosis (Osmosis is a process that involves water flow into a concentrated solution through a semipermeable membrane. The pressure by which water flows spontaneously is called as osmotic pressure. In reverse osmosis, impure water having dissolved salts is placed above a semipermeable membrane and subjected to high pressure. When the pressure is increased beyond the osmotic pressure, water flows in reverse direction. The pure water passes through the membrane and the salt is retained by the membrane), (f) Electro-dialysis (The high concentration of ionic impurities is removed by this process. The ions are transported through semi permeable membrane, influenced by an electric potential. The membranes are cation- or anion-selective, therefore either positive or negative ions are selectively passed through When. current is passed, cations get deposited on cathode and anions on the anode and the central compartment carries water which is free from ionic impurities).

7 Treatment Techniques in Mines

Following treatment techniques are commonly adopted at the mine sites for purifying the water: (a) Treatment of mine water discharge, (b) Treatment of effluent from the workshop and repairing and maintenance facilities, and (c) Treatment of the domestic sewage. The polluted water emanating from the mine is categorized into following three classes.

1. **Mine Effluent**

Mining industries generate a lot of effluent due to mining operations and below ground encounter water that is pumped out from the mine. Hence sedimentation tanks are constructed in such a way to settle down suspended solid sedimentation before being discharged.

2. **Workshop Effluent**

In major open-pit mines, a workshop effluent treatment plant (WETP) is constructed for washing and maintenance of vehicles. In this workshop effluents are generated, which needs to be treated in the effluent treatment plant. Basically, WETP were consisting of the pre-sedimentation tank, oil and grease trap, flash mixture, clarifloculator which clear water tank. In the year (2015) durgapur OCM (Chandrapur Area, WCL) open pit mine has reported recovery of oil and grease, and sludge of about 14 litres and 3.46 cubic meters for every month. Recovered materials were sent for processing in the third-party recovery plants [62].

3. **Domestic Effluent**

Untreated mining effluent is a serious issue for the environment. Water pollution from ground, surface, and acid mine leakage from mine waste has to be monitored as per Central Pollution Control Board (CPCB) guidelines and stipulation. The treatment is made through continuous aeration process based on the principle of stabilizing decomposable organic matter in the sewage, so as to produce an effluent and sludge [62].

8 The Role of Legislature in Mining Water Quality Management

The ever-increasing demand for water has led to shortage in mining areas and scarcity in many areas. Although, the concept of reduce, reuse, and recycle is already proliferated, a legal framework is necessary for management of water resources in mines. Therefore, the mining legislation was enforced for monitoring the use of groundwater for mining and beneficiation purposes. Although, the legislature mainly directs the use of groundwater, its subsidiary legislation provides technical guidelines by way of setting parameters for collection, coordination, interpretation, and storage of data. Such data can be modelled and used for making decisions, and to ensure that the drawdown of groundwater does not exceed the capacity of aquifers. An integrated water management in mining areas are largely governed by the extraneous, but essential influences, such as, political, societal, economic, and geopolitical. However, the water management and monitoring systems are often costly, and therefore out of favour by the mine owners. Therefore, the water rights and monitoring obligations remain largely unimplemented in the underdeveloped as well as developing countries [15, 64].

9 Water Quality Management in Mining Field

Optimum quality of water is essential to maintain for human and animal consumption, either directly through drinking or indirectly through food and crops. Issues of metal toxicity are severe in the active as well as derelict mines. Similarly, in absence of proper filtration mechanism, the mine water remains muddy, and often inundated by suspended particles and pathogens. Therefore, the management of water quality is very important in the mining fields. The water quality management is wholistic and encompasses the parameters such as geological and hydrogeological conditions of the area, availability and recharge of groundwater, optimum requirement and planning for judicious usage of groundwater, and prediction of short and long-term effects of water consumption. A variety of software such as, GoldSim™ are now available for mine water management [65].

10 Summary

Mining and subsidiary operations such as washing, beneficiation, and storage, have profound impact on the water quality. The direct exposure to hidden metallic as well as non-metallic constituents due to excavation strongly jeopardize the quality of water. The deteriorated water quality cannot be restored in the limited time period; therefore, it is important to make sure that the water being consumed is not hazardous for health. The water purification can be achieved by several ways, such as, implementing the water management systems. However, such systems are costly, therefore, barely implemented in the small mines and underdeveloped countries. However, the increased awareness about the quality of water can bring the desired change. Sustainability if mines is governed by the safety of mines, therefore, the issue of water quality in mines has not.

References

1. Chattopadhyay S, Chattopadhyay D (2012) Mining industries and their sustainable management. In: Meyers RA (ed) Encyclopedia of sustainability science and technology. Springer, New York, NY. https://doi.org/10.1007/978-1-4419-0851-3_864
2. Emmanuel AY, Jerry CS, Dzigbodi DA (2018) Review of environmental and health impacts of mining in ghana. J Health Pollut 8(17):43–52. https://doi.org/10.5696/2156-9614-8.17.43. Published 12 Mar 2018
3. Yadav HL, Jamal A (2015) Impact of mining on water resources in India. Int J Adv Res 3(10):1009–1015
4. Randive KR, Jawadand SA (2020) Mineral economics: an indian perspective. Nova Publishers, New York, p 320. ISBN: 978-1-53616-607-1
5. Overview of Mining and Mineral Industry in India (2001) Mining, minerals and sustainable development: tata energy research institute, p 94, [TERI Project Report No. 2001EE42]. https://pubs.iied.org/pdfs/G00615.pdf
6. The Indian mining sector effects on the environmental and FDI inflows (2002). https://www.oecd.org/env/1830307.pdf
7. Dhatrak SV, Nandi SS (2009) Risk assessment of chronic poisoning among Indian metallic miners. Indian J Occup Environ Med Aug 13(2):60–64. https://doi.org/10.4103/0019-5278.55121. PMID: 20386621; PMCID: PMC2847328
8. Das AP, Singh S (2011) Occupational health assessment of chromite toxicity among Indian miners. Indian J Occup Environ Med 15(1):6–13. https://doi.org/10.4103/0019-5278.82998. PMID: 21808494; PMCID: PMC3143520
9. Campaner VP, Luiz-Silva W, Machado W (2014) Geochemistry of acid mine drainage from a coal mining area and processes controlling metal attenuation in stream waters, southern Brazil. An Acad Bras Cienc 86(2):539–554. https://doi.org/10.1590/0001-37652014113712 Epub 2014 May 14 PMID: 24838545
10. Santiago PA, Iván PL, Susana LL, María VR (2014) Techniques to correct and prevent acid mine drainage: a review. DYNA 81(186):73–80. https://dx.doi.org/10.15446/dyna.v81n186.38436
11. Aznar-Sánchez JA, García-Gómez JJ, Velasco-Muñoz JF, Carretero-Gómez A (2018) Mining waste and its sustainable management: advances in worldwide research. Minerals 8:284
12. Sahni SK (2011) Hazardous metals and Minerals pollution in india: sources, toxicity and management. Indian Natl Sci Acad. http://www.insaindia.res.in/pdf/Hazardous_Metals.pdf

13. Kjellstrom T, Lodh M, McMichael T et al (2006) Air and water pollution: burden and strategies for control. In: Jamison DT, Breman JG, Measham AR et al (eds) Washington (DC): The international bank for reconstruction and development/the world bank. Chapter 43. Co-published by Oxford University Press, New York. https://www.ncbi.nlm.nih.gov/books/NBK11769/
14. Charbonnier (2001) Management of mining, quarrying and ore-processing waste in the European union study made for dg environment, European commission. December BRGM/RP-50319-FR
15. Mechlem K (2016) Groundwater governance: the role of legal frameworks at the local and national level—established practice and emerging trends. international natural resources law and human rights consultant, 69120 heidelberg, Germany. Water 8:347. https://doi.org/10.3390/w8080347
16. Mossa J, James LA (2013) Impacts of mining on geomorphic systems. Treatise on geomorphology. Academic Press, San Diego, CA, vol 13, pp 74–95. https://doi.org/10.1016/B978-0-12-374739-6.00344-4
17. Wuana RA, Okieimen FE (2011) Heavy metals in contaminated soils: a review of sources, chemistry, risks and best available strategies for remediation. Int Schol Res Notices. https://doi.org/10.5402/2011/402647
18. Soni AK (2019) Mining of minerals and groundwater in india. https://doi.org/10.5772/intechopen.85309
19. Sumi L, Thomsen S (2001) Mining in remote areas, issues and impacts. Minind Watch Canada/Mines Alerte
20. Karmakar HN, Das KP (2012) Impact of mining on ground & surface water. Int Mine Water Assoc Mining Quarry. http://parisara.kar.nic.in/PDF/Mining.pdf
21. Thimmaiah SA (2012) Socio-economics and environmental impact studies due to mining at bellary-hospetsector. Department of mining engineering national institute of technology Karnataka, Mangalore. https://idr.nitk.ac.in/jspui/bitstream/123456789/14455/1/050613MN05P01.pdf
22. Obasi PN, Akudinobi BB (2020) Potential health risk and levels of heavy metals in water resources of lead–zinc mining communities of Abakaliki, southeast Nigeria. Appl Water Sci 10:184. https://doi.org/10.1007/s13201-020-01233-z
23. Kaur P (2019) CAG report summary assessment of environmental impact due to mining activities and its mitigation in coal india limited. PRS Legislat Res. https://www.prsindia.org/report-summaries/assessment-environmental-impact-due-mining-activities-and-its-mitigation-coal-india
24. Chourasia LP (2007) Hydrogeochemistry of groundwater in a part ofthe hard rock terrain of central India. water quality and sediment behaviour of the future: predictions for the 21st century (proceedings of symposium HS2005 at IUGG2007, Perugia, July 2007). IAHS Publ. 314
25. Beeton RJS (Bob), Buckley Kristal I, Jones Gary J, Morgan Denise, Reichelt Russell E, Trewin D (2006) Australian state of the environment committee) 2006, Australia state of the environment 2006, independent report to the Australian government minister for the environment and heritage. Department of the Environment and Heritage, Canberra
26. Wen D, Zhang F, Zhang E, Wang C, Han S, Zheng Y (2013) Arsenic, fluoride and iodine in groundwater of China. J Geochem Explor 135:1–21
27. Winter TC, Judson W, Harvey O, Lehn F, Alley WM (1998) Ground water and surface water a single resource. U.S. Geological Survey Circular 1139. https://pubs.usgs.gov/circ/circ1139/pdf/circ1139.pdf
28. Fashola MO, Ngole-Jeme VM, Babalola OO (2016) Heavy metal pollution from gold mines: environmental effects and bacterial strategies for resistance. Int J Environ Res Public Health 2613(11):1047. https://doi.org/10.3390/ijerph13111047. PMID: 27792205; PMCID: PMC5129257
29. Khatri N, Tyag S (2014) Influences of natural and anthropogenic factors on surface and groundwater quality in rural and urban areas 23–39

30. Javed M, Usmani N (2019) An overview of the adverse effects of heavy metal contamination on fish health. Proc Natl Acad Sci India Sect B Biol Sci 89:389–403. https://doi.org/10.1007/s40011-017-0875-7
31. Hudson T (2012) Living with earth, an introduction to environmental geology. PHI Learning Private Limited
32. Rambabua K, Fawzi B, Pham QM, Shih-HsinHo N-Q, Loke P (2020) Biological remediation of acid mine drainage: review of past trends and current outlook. Environ Sci Ecotechnol 2:100024
33. Lecomte K, Maza S, Collo G (2017) Geochemical behavior of an acid drainage system: the case of the Amarillo River, Famatina (La Rioja, Argentina). Environ Sci Pollut Res 24:1630–1647. https://doi.org/10.1007/s11356-016-7940-2
34. Jennings SR, Neuman DR, Blicker PS (2008) Acid mine drainage and effects on fish health and ecology: a review. Reclamation Research Group Publication, Bozeman, MT, p 26
35. Yadav HL, Jamal A (2016) Treatment of acid mine drainage: a general review. Int Adv Res J Sci Eng Technol 3(11):116–122. https://doi.org/10.17148/IARJSET.2016.31123
36. Arnold DE (1991) Diversion wells—A low-cost approach to treatment of acid mine drainage. In: Proceedings 12th west virginia surface mine drainage task force symposium. Morgantown, WV
37. Aubé B (2004) Sludge disposal in mine workings at cape breton development corporation. MEND Ontario Workshop, Sudbury, pp 26–27
38. Aubé B, Zinck J (2009) Mine drainage treatment: a detailed review of active treatment options, advantages and challenges. short course presented at the 8th international conference on acid rock drainage
39. Watzlaf GR, Hyman DM (1995) Limitations of passive systems for the treatment of mine drainage. In: Proceedings 17th association of abandoned mine lands conference, pp 186–199
40. Yadav HL, Jamal A (2015) Removal of heavy metals from acid mine drainage: a review. Int J New Technol Sci Eng 2(3):77–84
41. Younger PL (2001) Passive treatment of ferruginous mine water using high surface area media. Elsevier science Ltd., Great Britain
42. Pescod MB (1992) Wastewater treatment and use in agriculture. FAO irrigation and drainage paper 47; Food and agriculture organization of the united nations, Rome. ISBN 92-5-103135-5
43. Thornton L, Butler D, Docx P, Hession M, Makropoulos C, McMullen M, Nieuwenhuijsen M, Pitman A, Rautiu R, Sawyer R, Smith S, White D, Wilderer P, Paris S, Marani D, Bragugli C, Palerm J (2001) Pollution in urban waste water and sewage sludge. ICON; I. C. Consultants Ltd., London, UK. European Commission, Final Report. ISBN 92-894-1735-8
44. Pinnarat N, Santirat N, Adisak P (2014) Risk assessment in the organization by using FMEA innovation: a literature review. proceedings of the 7th international conference on educational reform (ICER 2014), innovations and good practices in education: Global perspectives
45. Falkenmark M (1994) The dangerous spiral: near-future risks for water-related eco-conflicts. In: Proceedings of the ICRC symposium "water and war: symposium on water in armed conflicts", international committee of the red cross, Montreux, Switzerland, vol 16, pp 21–23
46. Faulkner BB, Skousen JG (1995) Effects of land reclamation and passive treatment systems on improving water quality. Green Lands 25(4):34–40
47. Benning BA, Otte ML (1997) Retention of metals and longevity of a wetland receiving mine leachate. In: Proceedings of 1997 national meeting of the american society for surface mining and reclamation, vol 16. Austin, TX, pp 43–46
48. Kleinmann RLP (1985) Treatment of acidic mine water by wetlands. USBM IC 9027:48–52
49. Skousen JG, Sexstone A, Ziemkiewicz PF (2000) Acid mine drainage control and treatment. https://doi.org/10.2134/agronmonogr41.c6
50. Jawadand SA, Randive KR (2021) A Sustainable approach to transforming mining waste into value-added products
51. Mohanty N, Goyal A (2012) Sustainable development: emerging issues in India's mineral sector. Planning commission. https://niti.gov.in/planningcommission.gov.in/docs/reports/sereport/ser/isid_mining%20_report1206.pdf

52. Development of Indian Mining Industry—The Way Forward. Non-Fuel Minerals. FICCI Mines and Metals Division October (2013). http://ficci.in/spdocument/20317/Mining-Industry.pdf
53. Rahmanian N, Hajar S, Ali B, Homayoonfard M. Ali NJ, Rehan M, Sadef Y, Nizami AS (2015) Analysis of physiochemical parameters to evaluate the drinking water quality in the state of Perak, Malaysia. J Chem 2015(716125):10. https://doi.org/10.1155/2015/716125
54. Shah CR (2017) Which physical, chemical and biological parameters of water determine its quality? Technical Report. https://doi.org/10.13140/rg.2.2.29178.90569
55. Tran Mien (2012) Mine waste water management and treatment in coal mines in Vietnam. Geosyst Eng 15(1):66–70. https://doi.org/10.1080/12269328.2012.674430
56. Nigam GK, Sahu RK, Sinha J, Sonwanshi RN (2015) A study on physico-chemical characteristics of water in opencast coal mine. J Ind Pollut Control 31(1):119–127
57. Rozkowski A, Rozkowski J (1994) Impact of mine waters on river water quality in the upper Silesian Coal Basin. Fifth International Mine Water Congress, Nottingham, 18–23 September, 811–821
58. Singh G (1994) Augmentation of underground pumped out water for potable purpose from coal mines of Jharia Coalfield. Fifth International Mine Water Congress, Nottingham, 18–23 September, 679–689
59. Hagare BD, Muttucumaru S, Singh RN (1995) Wastewater characteristics, management and reuse in mining and mineral processing industries. Water Res Risk Wastewater Recycle Reuse Reclam I:14–19
60. Kuyucak N (2001) Acid mine drainage treatment options for mining effluents. Mining Environ Manag
61. Khadsan RE, Kadu MV (2004) Drinking water quality analysis of some bore-wells water of chikhli Town, Maharashtra. J Ind Pollut Control
62. Pote NS (2017) Waste water treatment and management techniques in: mines. Inter J Eng Sci Res Tech 6(4):178–192. https://doi.org/10.5281/zenodo.557144
63. Yang PY, Zhang ZQ (1995) Nitrification and denitrification in the wastewater treatment system. In: Proceedings of the UNESCO—University of tsukuba international seminar on traditional technology for environmental conservation and sustainable development in the asian-pacific region. Tsukuba Science City, Japan
64. Tripathi A, Mishra AK, Verma G (2016) Impact of preservation of subsoil water act on groundwater depletion: the case of Punjab India. Environ Manag 58:48–59
65. Punkkinen H, Räsänen L, Mroueh UM, Korkealaakso J, Luoma S, Kaipainen T, Backnäs S, Turunen K, Hentinen K, Pasanen A, Kauppi S, Vehviläinen B, Krogerus K (2016) Guidelines for mine water management. VTT Technology 266. http://urn.fi/URN:ISBN:978-951-38-8443-7

Mineral and Mining Wastes: A Burgeoning Problem with a Need for Sustainable Restitution

Nikhil P. Kulkarni

Abstract Mining produces substantial amount of wastes, the quantum of which is directly linked to the volume of mining. Whilst generation of this high-volume waste is initiated from the very first phase of extraction, various forms of it continue to be generated during the subsequent phases, right until extraction and beneficiation of minerals. Comprising chiefly of overburden material, mine water, slurry and tailings, the composition of these wastes is entirely dependent on the type of mine being exploited. Waste rock or overburden is the rock material that needs to be removed in order to reach the economic mineral of interest. Its quantity therefore depends on the mining technique adopted as well as the location of the ore body in the earth's crust. Mine water, also a waste resulting from mining activities, is a reservoir of potentially contaminating chemicals that tend to concentrate, as a result of increased percolation, with an increase in mining activities. Slurry and tailings tend to be formed post processing of the mined ore during the mineral extraction process, the disposal of which today is one of the biggest environmental concern. Usually fine grained and mud like, they comprise ground rock and process effluents which are mostly toxic and which essentially require chemical treatment prior to disposal. Developments in engineering and technology have today opened up newer ways for disposal or alternate uses of these wastes mostly by rendering them inert. What however remains is the strategic and financial will to adopt these novel technologies towards a sustainable future.

1 Introduction

Extraction of mineral resources is known to provide appreciable opportunities for economic development thereby serving as a catalyst for employment, export earnings, infrastructure development and increased tax revenues. The World Bank has been known to have supported forty-one mining sector reform projects in twenty-four countries since 1988 which have in turn led to an increase in investments coupled

N. P. Kulkarni (✉)
National Institute of Miners' Health, Nagpur, India

K. Randive et al. (eds.), *Innovations in Sustainable Mining*, Earth and Environmental Sciences Library, https://doi.org/10.1007/978-3-030-73796-2_13

with other related economic indicators such as Gross Domestic Product (GDP), fiscal revenues, etc. The World Bank's mining sector is also known to promote policies and programs that fortify governance without affecting the environment thereby ensuring that the benefits reaped are widespread and sustained [1].

A representative and provisional graph highlighting the world total coal production as per International Energy Agency (IEA) records for the last forty years [2] is as shown in Fig. 1 below. The increased coal production worldwide and therefore also the energy trend is but evident.

Mining involves extraction of geological resources of economic value from deposits in the Earth's crust. This is bound to affect the environment by inducing loss of biodiversity, soil erosion and contamination of water resources. Other than the visually evident environmental damages, mining over a prolonged period of time is known to cause leakage and leaching of chemicals from the mining sites which can have damaging effects on the health of the population living in the periphery of the mining site. And despite the many environmental legislations in effect towards rehabilitation of mined sites, violation of these is but common [3]. Collectively addressed to as "mining wastes", these comprise high volumes of material, the origination of which begins right from excavation viz. overburden, and continues right

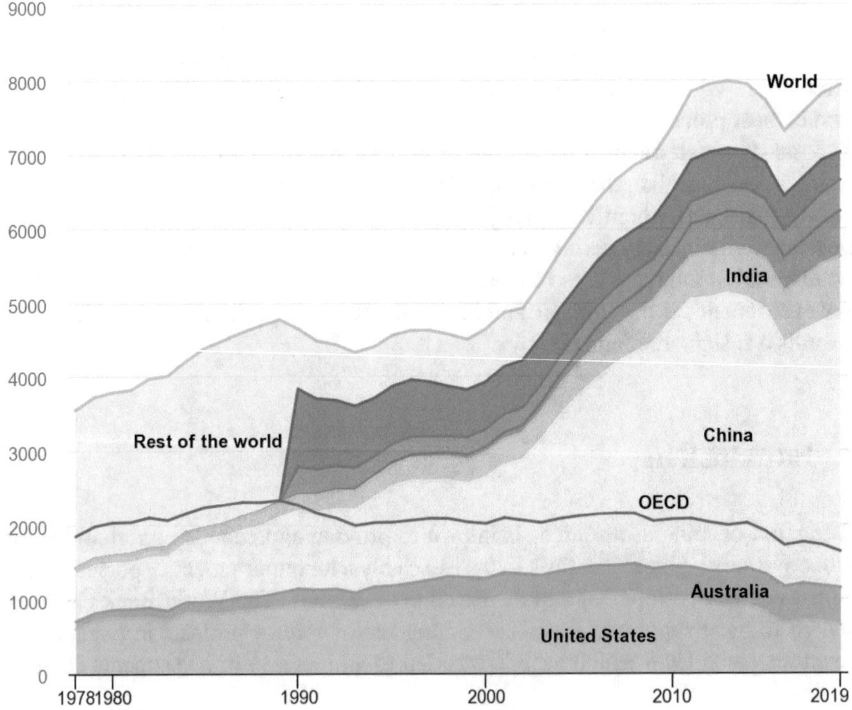

Fig. 1 World total coal production (provisional 1971–2019)

Table 1 Generation of waste from processing of metal ores

Metal	Typical percentage of metal in ore (%)	Percentage waste	Waste mineral ratio
Iron	33.00	60	5.2:1
Copper	0.6	99	450:1
Gold	0.0004	99.99	950000:1
Lead-zinc	3.7–5.0	97.5	32:1
Aluminium	8–20	70	3:1
Silver	0.03	99.97	NA
Tungsten	0.5	99.5	NA
Molybdenum	0.2	99.8	NA

up to discarded materials such as slurry and tailings which emanate post processing of the mined ore during the mineral extraction process.

Literature available and depicting the volume of waste generated for a few select ores [4] is as given in Table 1 below.

In a broader sense it can therefore be said that the kind and quantity of mining waste, as also its share in the total waste stream is highly dependent on the availability of the natural resources of the country, and the economic value of the mineral in addition to its market demand at a given time [5].

2 Overburden

Broadly speaking, overburden in the mining terminology refers to the material which needs to be removed so as to gain access to the ore body for mining and/or economical exploitation [6, 7]. Usually comprising of soil, rock, and an ecosystem that lies above an ore body, overburden needs to be typically removed for surface mining and requires concurrent planning along with mining activities and site rehabilitation [8, 9]. As mine development usually progresses, the removed topsoil and rocks are stored in dumps around the mining operation area [10] with an aim to either use them for reclamation or land filling [11].

Under normal conditions these are not contaminated with toxic components [12] and may or may not be voluminous. However, the amount of overburden removed and its excavation rate are usually observed to be related to

- required ore productivity,
- morphology and geometry of the ore deposit and
- the effectiveness of the disposal processes [13].

Over 90% of the mining activities in the developed countries are known to adopt open pit methods [14] wherein the ore body is usually relatively close to the surface [13]. But with the lure of productivity, overburden disposal often gets the cold shoulder until not made mandatory.

Being the most abundant fossil fuel on Earth [15], coal mines usually make up for the major chunk of mining in the world and are a major source of fuel for electricity generation. Therefore, not just do they contribute big time to the social and economic benefits, and therefore the GDP of a country but their share in deterioration and degradation of the environment too, is on a similarly large scale [16, 17]. Sufficient research has proven the link coal mining has with adverse effects on human health [18], ecological degradation, ill effects on environmental landscape and socio-cultural-economic conditions [19, 20] as a result of deforestation, land subsidence, coal dust generation, mine waste generation, spontaneous fires, etc. [21]. Accumulation of overburden has also been proven to be linked to a prolonged removal of forest cover and detriment of the top soil. Overburden of coal mines also often referred to as coal mine spoils along with the waste rocks is known to spontaneously oxidise in the open atmosphere. This causes release of metal rich effluents to the surrounding environs and the neighbouring water bodies [22–24]. Soils are known for their metal scavenging potential, and therefore also act as a metal sink for the trace elements released [25]. However, at the cost of eroding the chemical quality and the microbiological fauna of the soil [26–30] which in turn pose a threat to the humans and the ecosystem as they (trace metals) enter the food chain [28, 29, 31]. However, it is imperative to note that the average concentration of trace elements such as arsenic, cadmium, copper, chromium, mercury, manganese, nickel, lead and zinc and its resultant metal pollution near coal mines when compared to world background soils is "site specific" [32]. Bioaccumulation of trace metals specific to the metalliferous mine spoils has been reported with its elevated levels observed in the above-ground parts of vegetation [33] in its vicinity.

Further, one of the chief inorganic components of coal is pyrite (FeS_2) [34, 35], wherein the concentration of sulphur ranges between 2–11% [36]. These elevated levels of sulphur which therefore reflected in the accumulated coal mine spoils turn the metal leachates acidic with pH ranging between 2.0–3.0 [35, 37]. The contamination of the surface and sub-surface water bodies [38] by the leachates is but self-explanatory.

In addition to pollution of water resources, excavation of top soil for coal recovery also gives rise to air pollution in the nearby vicinity as the loose soil from the over burden and coal dust from the open mine get airborne [39–42]. Hence what usually starts out visually only as excavation and deforestation for development of a mine, usually and mostly turns to disruption of water regimes, bioaccumulation in food chains, etc. on the micro levels as overburden or mine spoils accumulate.

3 Acid Mine Drainage

Sub-surface mining often advances to levels below the water table, thereby necessitating the need to pump water out so as to prevent inundating the mines with water. However, post mining when mines are abandoned, pumping is discontinued following which water floods the mine. It is situations like these which form the prefatory for most of the acid mine drainage situations [43].Whilst there have been situations wherein drainage, an unavoidable product, from mines wasn't acidic or where the acidity of the mine drainage got neutralized over the flow path over a period of time thereby rendering it neutral or otherwise, but prefixes other than "acidic" for mine drainage never gained acceptance following which "**A**cid **M**ine **D**rainage" (AMD) became the term of choice [44, 45]. The causative sources for AMD are usually mine waste rock dumps [44] and mine (mostly coal) spoils, collectively referred to as overburden and the tailing ponds [46].

Similar to the niche of coal mining in the mining industry, AMD from coal mines too has its own exclusive nomenclature–**C**oal **M**ine **D**rainage (CMD).

A pictorial depiction of CMD genesis from coal mines is as shown below in Fig. 2.

Characterised by an increased concentration of heavy metal sulphates as a result of pyrite presence in coal and its spoils, CMD usually exhibits low pH and high electrical conductivity [47], with the heavy metals usually being arsenic, iron, lead, cadmium, zinc, mercury, etc. [48, 49]. Pyrite from the spoils on leaching through the strata forms an acidic drainage due to the metabolic activity of acidophilic bacteria, *Thiobacillus ferroxidans* which is known for being able to catalyse pyrite oxidation [19], the precursor to CMD formation. A simultaneous reaction induced by weathering of rocks combined with anthropogenic factors in coal spoils leads to the formation of dilute sulphuric acid which is another contributory agent in CMD generation. As CMD accumulates, so do levels of conductivity, TDS, TSS, oil, grease, sulphates, calcium, magnesium and iron increase thereby decreasing its potability before it

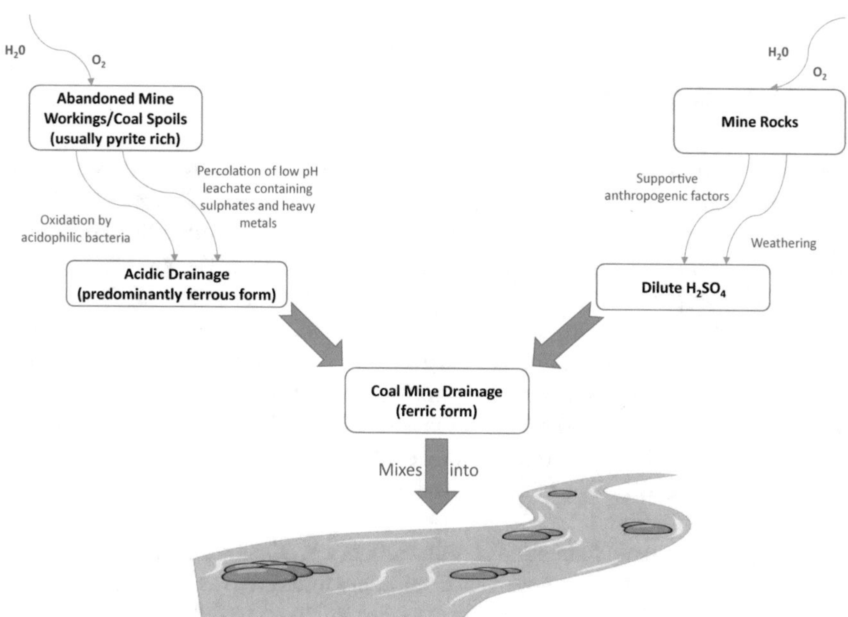

Fig. 2 Generation of CMD from coal mines

finally enters into water bodies, streams and rivers. This untreated CMD degrades the water quality making it unfit for use and at the same time even affects the aquatic life in a chronic sense [50]. Further, presence of non-degradable heavy metals in the CMD, make its presence felt in the water bodies and over a period of time enter the food chain and thereby have an adverse and long-lasting effect on the plants or animals leading to a detriment in biological functions [51]. Research on CMD has also proven that elevated levels of hydroxides in them, especially the iron hydroxides, physically coat the surface of stream beds thereby destroying aquatic habitat and fish food as well as diminishing the availability of clean gravels necessary for spawning [20, 52]. However, it is also important to note that the severity and extent of water body deterioration is largely dependent on its buffering capacity, as also the chemistry of the CMD along with its volume and frequency [53].

In the case of AMD arising from limestone mines, classification is chiefly based on iron content, with AMD classified into three types, viz.—low iron AMD, ferric iron AMD and ferrous iron AMD. The mechanism of action followed by the respective AMD in effecting pollution is the same as that of the CMDs. However, the former is observed to be easier to treat as a result of its less acidic pH, whereas the latter is considered the most difficult, due to increased reaction times with treatment agents [54].

In a wider sense, while occurrence of CMD has been well documented, research is however scanty for AMD resulting from non-coal and metal mines, despite elevated levels of respective metals from each of these mines posing a threat as serious as CMD.

4 Slurry and Tailings

With composition being similar to that of the parent ore in addition to processing chemicals, mine tailings are basically fine-grained particles left behind post ore processing carried out for removal of minerals with economic value. Whilst the coarse contents usually consisting of sand and gravel constitute 15–50% of the tailings, the quantity of fine particles (viz. clay and silt) suspended in water [55] is inversely proportional to grade of ore being mined. Slurry and tailings have been reported as having the potential to scathe the environment by polluting it with toxic metals (arsenic and mercury being two major culprits), radioactive material, etc. [10]. These can in turn cause erosion and sink holes in addition to polluting groundwater, rivers, soil and air [56] which in turn irreversibly contaminate the food chain.

Typical particle size distribution curves of tailings [57] from varied mines are as illustrated below in Fig. 3.

Tailings from bauxite mines and alumina refinery residues are even known to have a unique nomenclature for themselves. Commonly referred to as Red Mud due to its colour, as a result of the presence of iron oxides, often up to 60%, and its very fine particle size, these tailings are generated as a result of the Bayer's process commonly applied in refineries for refining bauxite ore into alumina which is further used to

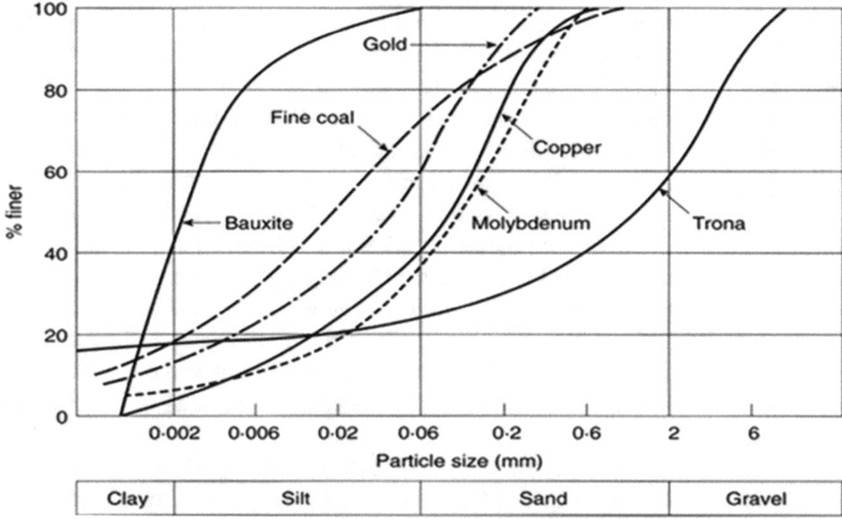

Fig. 3 Particle size distribution curves for tailings from some mines

produce aluminium [58]. A typical bauxite refinery is said to produce almost one to two times of red mud as against the amount of alumina produced. Highly alkaline in nature, with pH in the range of 10–13, red mud is known to cause environmental concerns as a result of its high alkalinity as well as storage problems due to its fine size and voluminous generation [59]. Significant amount of research is therefore being indulged in throughout the world for efficiently dealing with red mud so as to amalgamate it for practical utility.

Another ore, the tailings of which have caused similar environmental furore is that of copper wherein estimates suggest generation of solely one hundred and twenty-five tons of tailings per one ton of produced copper [60] based on the grade of ore being mined. These tailings are known to contain reactive sulphide minerals and residual metal minerals [61], indicating the need for systematic and safe storage in tailing dams failing which rapid oxidation can lead to acidic water production which in turn can carry with it toxic metals such as arsenic and mercury to the environment in the event of percolation [62] in case of tailing-dam failure. Developing and exploiting technological know-how for monitoring the associated risk factors is perhaps therefore the only safeguard.

Being fine grained due to high silt concentrations, slurry and tailings also tend to have a high surface area, thereby being susceptible to physical and chemical changes, especially atmospheric oxidation [63]. And this in turn necessitates proper storage measures to forestall oxygen-mediated chemical changes, especially tailings with sulphide content [64] which are often referred to as the largest environmental liability of the mining industry [65].

Mine tailings are usually stored in storage facilities which are isolated impoundments. Whilst some are created under water, impoundments or earth filled embankment dams known as tailing dams too are sometimes created wherein the material used to raise the dam height over the life of a particular mine often includes tailings along with dirt [66]. However, it is very important to ascertain the behaviour of the tailings proposed to be deposited in the impoundments with respect to their potential short- and long-term liabilities. Design of a tailing dam for its optimal operational efficiency should be carried out only after the said characteristics have been established; as an improper blueprint can negate the very purpose the tailing dam was created for. History of tailing dam failures has been witness to unstable tailing piles causing landslides, dry tailing deposits getting airborne to create health problems in nearby communities and rains resulting in leaching wherein acidic water leakage disrupted aquatic life downstream or radioactivity levels peaking in surrounding areas [67].

After effects post the toxic waste episode at the McArthur River zinc and lead deposit at Borroloola on the Gulf of Carpentaria in Australia involving creek and fish contamination respectively is but a glaring example of environment degradation post tailing seepage from its Tailings Storage Facility [68] and suspected leaching of AWD into water resources [69]. One of the most recent disasters involving collapse of a tailing dam at the Córrego do Feijão iron ore mine in Brazil in early 2019 involved the death of 272 people. Classified as a catastrophic failure, the Brumadinho dam disaster happened after release of a mudflow totalling 12 million cubic meters of tailings that was capable of polluting over three hundred kilometres of the river passing through Parque de Cachoeira post dam collapse and which advanced through mine offices as well as its cafeteria during lunch hours with further advancement proceeding downstream to houses, farms and roads [70].

Studies have established that the highest proportion of tailing dam collapses (>50%) were those built with the upstream construction method whereas water retention dam models accounted for the next highest proportion [71]. The need for comprehensive evaluation of geological conditions and environmental factors prior to construction of tailing dams is therefore highlighted since these build up dynamically over time as mine wastes pile up. Construction of tailing dams using the downstream construction model coupled with centreline construction method has therefore been advocated post assessment of geological weaknesses in the ground below. Incorporation of a reliable flood discharge system in the tailing dam can prevent collapses that happen as a result of stress and liquefied tailings.

Other than negligence and human errors, tailing dams are also at a risk from earthquakes or undiscovered geologic fault-lines, which carry with it a huge factor of unpredictability. For tailing dams in earthquake prone areas, reinforced foundations can help minimize detrimental effects of a tailing dam collapse. Additionally, incorporation of technology that uses radars or lasers to detect worrisome bulges can also be used to detect signs of tailing dam stress or failure [72].

In contrast with disposal of overburden and tailings from mineral ore and coal mines, the disposal of waste from radioactive mines and allied millings needs incorporation of additional measures, the guidelines for which have been specified by

International Atomic Energy Agency (IAEA), Vienna [73]. Herein, post classifying the wastes based on their level of radioactivity, IAEA recommends in situ stabilization followed by disposal involving isolation and burial underground.

5 Sustainable Management of Mining Waste—The Need of the Hour

Despite the pressing need, sustainable redressal of mining waste has rarely received the importance it deserves. It can also not be negated that the approach to sustainability with regards to mining and its waste requires a multipronged approach combining availability of finances with technical knowhow, available materials and natural resources; and the same can be achieved only through significant co-operation and contribution of all involved stakeholders [74].

With the leaps and bounds of technical and technological know-how available today, it is possible to ascertain behaviour and reactivity of mining waste, predicting its transport potential using modelling tools and assessing stability and risks associated with dams and tailing ponds constructed for mining waste disposal [75]. Application of these tools can be included in pre-approval sanctions needed prior to mining waste disposal.

Amidst increasing awareness of repercussions due to irresponsible mining and/or mining waste handling there is today an urgent need to further research along the lines of sustainable mining methods and de-polluting mining waste. A well spelt out Mine Closure Plan can play an important role in sustainable development post mining. Mine closure plans usually involve two components—progressive and final closure. While the progressive closure plans involve land use/reclamation activities carried out continuously and sequentially during the mining operation, the final mine closure activities are implemented towards exhaustion of reserves and mine closedown and continue until the mine area is restored to environmentally acceptable standards or better [76].

In accordance with the stage of mine closure, overburden dump stability can be attempted to be enhanced using Miyawaki method of afforestation or other revegetation programmes involving coir matting, vetiver grass, etc. [77]. Use of solid mining waste residues as components of building materials, and, mining slag and gases for electricity generation are other current key research thrusts in an attempt to remediate and reuse mining waste. Copper tailings, considered an environmental burden due to its volume, can be utilised as a source of low-cost mineral resources since comminution is not required to process them [78]. Additionally, studies have also demonstrated their use in production of paving stones which can be used in the construction sector [79]. Similarly, iron ore mine tailings [80] and gold mill tailings [81] too have been reported to be used for geo-polymer and soil tailing bricks manufacture respectively. In lines with a sustainable approach towards mining, coal overburden material has oft been used for stowing operations since it not just helps prevent land subsidence but also helps reduce mine fires along with aiding in improved extraction of coal pillars thereby meliorating coal production [82].

Water contamination is often, if not always, the first visual indicator of improper mining waste disposal. Amplification of the said contamination can be attempted to be resolved via usage of nutrient enriched sediments so as to reduce acidity in addition to applying engineering systems for its storage. Other lines of research along similar lines towards AMD remediation currently being pursued involve use of microbes and biotechnology, phyto-remediation, bioremediation with algae etc.

Adoption of sustainable practices will therefore be the only buzzword for the future.

References

1. World Bank Group (2013) Mining: sector results profile. https://www.worldbank.org/en/results/2013/04/14/mining-results-profile. Accessed 02 Nov 2019
2. International Energy Agency (2020) World total coal production, 1971–2019 provisional. https://www.iea.org/data-and-statistics/charts/world-total-coal-production-1971-2019-provisional. Accessed 10 Dec 2020
3. Chepkemoi J (2017) What is the environmental impact of the mining industry? World Atlas. https://www.worldatlas.com/articles/what-is-the-environmental-impact-of-the-mining-industry.html. Accessed 02 Nov 2019
4. Pal BK, Khanda DK, Dey S (2012) Problems of mining wastes management in India and its suggestive measures–case studies. In: Paper presented at the 5th international congress of environmental research. National Institute of Technology, Rourkela, pp 22–24
5. Szczepanska J, Twardowska I (2004) Mining Waste. In: Chapter III. 6, Waste management series, vol 4. Elsevier, pp 319–385
6. Rankin WJ (ed) (2011) Minerals, metals and sustainability: meeting future material needs. CSIRO publishing, Australia
7. Vela-Almeida D, Brooks G, Kosoy N (2015) Setting the limits to extraction: a biophysical approach to mining activities. Ecol Econ 119:189–196
8. Allsman PT, Yopes PF (1973) Open-pit and strip-mining systems and equipment. In: Cummins AB, Given IA (eds) Mining engineering handbook, vol 2, section 17. SME Mining, New York
9. Galera JM, Checa M et al (2009) Enhanced characterization of a soft marl formation using in situ and lab tests for the prestripping phase of Cobre Las Cruces open pit mine. Slope Stability, Santiago de Chile
10. Franks DM, Boger DV, Côte CM et al (2011) Sustainable development principles for the disposal of mining and mineral processing wastes. Res Policy 36(2):114–122
11. Sheoran V, Sheoran AS, Poonia P (2010) Soil reclamation of abandoned mine land by revegetation: a review. Intl J Soil Sedim Water 3(2):13
12. Kogel JE, Trivedi NC, Barker JM, Krukowski ST (eds) (2006) Industrial minerals & rocks: commodities, markets, and uses. Soc Mining Metall Explor, Inc. Colorado
13. Oggeri C, Fenoglio TM, Godio A et al (2019) Overburden management in open pits: options and limits in large limestone quarries. Intl J Mining Sci Technol 29(2):217–228
14. Mossa J, James LA (2013) Impacts of Mining on Geomorphic Systems. In: James LA, Harden C, Clague J (eds) Geomorphology of human disturbances, climate change, and natural hazards. Elsevier, Treatise on Geomorphology, pp 74–95
15. Elliott MA (1981) Chemistry of coal utilization, vol second supplementary. John Wiley and Sons, New York

16. Hower JC, Graham UM, Wong AS et al (1998) Influence of flue-gas desulfurization systems on coal combustion by-product quality at Kentucky power stations burning high-sulfur coal. Waste Manag 17(8):523–533
17. Tiwary RK (2001) Environmental impact of coal mining on water regime and its management. Water Air Soil Pollut 132(1–2):185–199
18. Sahoo D, Bhattacharjee A (2010) An Assessment of environmental impact of coalmines on the local inhabitants: a case study of north eastern coalfields, Margherita, Assam. Indian J Econ Bus 9(4):841
19. Baruah BP, Khare P (2010) Mobility of trace and potentially harmful elements in the environment from high sulfur Indian coal mines. Appl Geochem 25(11):1621–1631
20. Giri K, Mishra G, Pandey S et al (2014) Ecological degradation in northeastern coal fields: Margherita Assam. Int J Sci Environ Technol 3(3):881–884
21. Equeenuddin SK Md. (2010) Controls of coal and overburden on acid mine drainage and metal mobilization at makum coalfield, Assam, India. PhD Thesis, IIT Kharagpur, India
22. Equeenuddin SM, Tripathy S, Sahoo PK et al (2010) Hydrogeochemical characteristics of acid mine drainage and water pollution at Makum Coalfield India. J Geochem Explor 105(3):75–82
23. Chuncai Z, Guijian L, Dun W et al (2014) Mobility behavior and environmental implications of trace elements associated with coal gangue: a case study at the huainan coalfield in China. Chemosphere 95:193–199
24. Sahoo PK, Tripathy S, Equeenuddin SM et al (2012) Geochemical characteristics of coal mine discharge vis-à-vis behavior of rare earth elements at Jaintia Hills coalfield, northeastern India. J Geochem Explor 112:235–243
25. Banat KM, Howari FM, Al-Hamad AA (2005) Heavy metals in urban soils of central Jordan: should we worry about their environmental risks? Environ Res 97(3):258–273
26. Bhuiyan MA, Parvez L, Islam MA et al (2010) Heavy metal pollution of coal mine-affected agricultural soils in the northern part of Bangladesh. J Hazard Mater 173(1–3):384–392
27. Sahoo PK, Bhattacharyya P, Tripathy S et al (2010) Influence of different forms of acidities on soil microbiological properties and enzyme activities at an acid mine drainage contaminated site. J Hazard Mater 179(1–3):966–975
28. Wuana RA, Okieimen FE (2011) Heavy metals in contaminated soils: a review of sources, chemistry, risks and best available strategies for remediation. Isrn Ecol. https://doi.org/10.5402/2011/402647
29. Zhang Z, Abuduwaili J, Jiang F (2013) Determination of occurrence characteristics of heavy metals in soil and water environments in Tianshan Mountains Central Asia. Anal Lett 46(13):2122–2131
30. Harun-Or-Rashid SM, Roy DR, Hossain MS et al (2014) Impact of coal mining on soil, water and agricultural crop production: a cross-sectional study on Barapukuria coal mine industry Dinajpur Bangladesh. J Environ Sci Res 1(1):1–6
31. Nagajyoti PC, Lee KD, Sreekanth TVM (2010) Heavy metals, occurrence and toxicity for plants: a review. Environ Chem Lett 8(3):199–216
32. Sahoo PK, Equeenuddin SM, Powell MA (2016) Trace elements in soils around coal mines: current scenario, impact and available techniques for management. Current Pollut Rep 2(1):1–14
33. Xiao WL, Luo CL, Chen YH et al (2008) Bioaccumulation of heavy metals by wild plants growing on copper mine spoils in China. Commun Soil Sci Plant Anal 39(3–4):315–328
34. Monterroso C, Macías F (1998) Prediction of the acid generating potential of coal mining spoils. Int J Surf Mining Reclam Environ 12(1):5–9
35. Baruah BP, Saikia BK, Kotoky P et al (2006) Aqueous leaching on high sulfur sub-bituminous coals, in Assam India. Energy Fuels 20(4):1550–1555
36. Chabukdhara M, Singh OP (2016) Coal mining in northeast India: an overview of environmental issues and treatment approaches. Int J Coal Sci Technol 3(2):87–96
37. Dowarah J, Boruah HD, Gogoi J et al (2009) Eco-restoration of a high-sulphur coal mine overburden dumping site in northeast India: a case study. J Earth Syst Sci 118(5):597–608

38. Sheoran V, Sheoran AS, Tholia NK (2011) Acid mine drainage: an overview of Indian mining industry. Int J Earth Sci Eng 4(6):1075–1086
39. Dutta P, Mahatha S, De P (2004) A methodology for cumulative impact assessment of opencast mining projects with special reference to air quality assessment. Impact Assess Project App 22(3):235–250
40. Dkhar AA, Rai RK (2005) Impact of Coal Mining on Micro-landforms in Jaintia Hills District, Meghalaya. In: Singh OP (ed) Mining environment: problems & remedies. Regency Publishing, New Delhi, pp 41–56
41. Borpujari D, Saikia LR (2006) A study of growth performance of five dominant plant species in coal mine spoil at Tikak opencast mine under the Patkai range of eastern Himalaya. Nature Environ Pollut Technol 5(1):13–20
42. Sarma K, Barik SK (2011) Coal mining impact on vegetation of the Nokrek biosphere reserve, Meghalaya India. Biodiversity 12(3):154–164
43. Wikipedia. https://en.wikipedia.org/wiki/Acid_mine_drainage. Accessed 26 Nov 2019
44. The International Network for Acid Prevention. https://www.inap.com.au. Accessed 28 Nov 2019
45. Gusek JJ, Wildeman TR, Conroy KW (2006) Conceptual methods for recovering metal resources from passive treatment systems. In: Proceedings of the 7th international conference on acid rock drainage, St. Louis MO, pp 26–30
46. Marquez JE, Pourret O, Faucon MP et al (2018) Effect of cadmium, copper and lead on the growth of rice in the coal mining region of Quang Ninh, Cam-Pha (Vietnam). Sustainability 10(6):1758
47. Somerset VS, Petrik LF, White RA et al (2005) Alkaline hydrothermal zeolites synthesized from high SiO_2 and Al_2O_3 co-disposal fly ash filtrates. Fuel 84(18):2324–2329
48. Guha Roy PK (1991) Coal mining in Meghalaya and its impact on environment. Exposure 4:31–33
49. Singh PK, Singh AL, Kumar A et al (2012) Mixed bacterial consortium as an emerging tool to remove hazardous trace metals from coal. Fuel 102:227–230
50. Dutta M, Saikia J, Taffarel SR et al (2017) Environmental assessment and nano-mineralogical characterization of coal, overburden and sediment from Indian coal mining acid drainage. Geosci Front 8(6):1285–1297
51. Järup L (2003) Hazards of heavy metal contamination. British Med Bull 68(1):167–182
52. Hoehn RC, Sizemore DR (1977) Acid Mine Drainage (AMD) and its impact on a small Virginia stream. J Am Water Res Assoc 13(1):153–160
53. Kimmel WG (1983) The impact of acid mine drainage on the stream ecosystem. In: Miller WW, Majumdar SK (eds) Pennsylvania coal: resources, technology, and utilization. Pennsylvania Academic Science Publications, USA, pp 424–437
54. Metallurgist Corp (2019) Limestone treatment of acid mine drainage. https://www.911metallurgist.com. Accessed 09 Dec 2019
55. Vick SG (1990) Planning, design, and analysis of tailings dams. BiTech Publishers Ltd., Vancouver
56. Murdoch University (2017) Dangers of Mine Waste highlighted in U.N. report. https://phys.org. Accessed 10 Dec 2019
57. Sarsby RW (2000) Environmental geotechnics. Thomas Telford, London
58. Schmitz C (ed) (2006) Handbook of aluminium recycling. Vulkan Verlag, Germany
59. Evans K (2016) The history, challenges, and new developments in the management and use of bauxite residue. J Sustain Metall 2(4):316–331
60. Cheng TC, Kassimi F, Zinck JM (2016) A holistic approach of green mining innovation in tailings reprocessing and repurposing. In: Proceedings of Tailings and Mine Waste, Keystone. Colorado, USA, pp 2–5
61. Beauchemin S, Clemente JS, Thibault Y et al (2018) Geochemical stability of acid-generating pyrrhotite tailings 4 to 5 years after addition of oxygen-consuming organic covers. Sci Total Environ 645:1643–1655

62. Shepherd T, Rumengan I, Sahami A (2018) Post-depositional behaviour of mercury and arsenic in submarine mine tailings deposited in Buyat Bay, North Sulawesi, Indonesia. Marine Environ Res 137:88–97
63. Kossoff D, Dubbin WE, Alfredsson M et al (2014) Mine tailings dams: Characteristics, failure, environmental impacts, and remediation. Appl Geochem 51:229–245
64. Geological Survey of Sweden (2019) Lecture 4: Mining Waste. https://www.sgu.se/en/geointro. Accessed 09 Dec 2019
65. Nehdi M, Tariq A (2007) Stabilization of sulphidic mine tailings for prevention of metal release and acid drainage using cementitious materials: a review. J Environ Eng Sci 6(4):423–436
66. Blight GE (1998) Construction of Tailings Dams. Case studies on tailings management. International Council on Metals and the Environment, Paris, France, pp 9–10
67. Treehugger (2018) Mine tailings and the environment. https://www.thoughtco.com. Accessed 14 Dec 2019
68. Northern Territory Government (2011) Independent monitor's audit of the McArthur river mine for the 2010 operational period. Report To The Minister For Primary Industry Fisheries and Resources, Australia
69. Ltd Indo-Pacific Environmental Pty (2015) Monitoring of metals and lead isotope ratios in fish, crustaceans and molluscs of the McArthur River. McArthur River Mine, Australia
70. Wikipedia. https://en.wikipedia.org/wiki/Brumadinho_dam_disaster. Accessed 10 Dec 2019
71. Witt KJ, Schönhardt M (eds) (2004) Report: Tailings management facilities–risks and reliability. Tailsafe
72. Lyu Z, Chai J, Xu Z et al (2019) A comprehensive review on reasons for tailings dam failures based on case history. Adv Civil Eng 2019:1–18. https://doi.org/10.1155/2019/4159306
73. International Atomic Energy Agency (2011) Disposal of radioactive waste, iaea safety standards series No. SSR 5. Vienna
74. Aznar-Sánchez JA, García-Gómez JJ, Velasco-Muñoz JF et al (2018) Mining waste and its sustainable management: advances in worldwide research. Minerals 8(7):284
75. Charbonnier P (ed) (2001) Report: management of mining, quarrying and ore-processing waste in the European Union. Bureau de Recherches Geologiques et Minieres
76. Ministry of Coal, Government of India (2013) Mine closure guidelines. India
77. Ranjan V, Sen P, Kumar D et al (2015) A review on dump slope stabilization by revegetation with reference to indigenous plant. Ecol Proc 4(1):1–11
78. Santander M, Valderrama L (2019) Recovery of pyrite from copper tailings by flotation. J Mater Res Technol 8(5):4312–4317
79. Lam EJ, Zetola V, Ramírez Y et al (2020) Making paving stones from copper mine tailings as aggregates. Int J Environ Res Public Health 17(7):2448
80. Kuranchie FA, Shukla SK, Habibi D (2016) Utilisation of iron ore mine tailings for the production of geopolymer bricks. Int J Mining Reclam Environ 30(2):92–114
81. Roy S, Adhikari GR, Gupta RN (2007) Use of gold mill tailings in making bricks: a feasibility study. Waste Manag Res 25(5):475–482
82. Gupta AK, Paul B (2015) A review on utilisation of coal mine overburden dump waste as underground mine filling material: a sustainable approach of mining. Int J Mining Miner Eng 6(2):172–186

Distinctive Bats Species in Abandoned Mines: Adventure Geotourism for Nature Enthusiasts

Madhuri Thakare and Kirtikumar Randive

Abstract Abandoned mines had never been the famous and popular tourist destinations, because of difficulty of access, non-maintenance of sites, waste water ponds which are often filled with unhealthy and contaminated waters, fragile waste dumps, anthropogenic wastes often containing sharp metallic tools such as nails, rods, machine parts, abandoned vehicles, plastic wastes such as containers, bottles, and, others. Moreover, such mines are often inhabited by wildlife, bats, snakes, worms, and insects. There are many other possible dangers associated with the abandoned underground mines, namely, fragile roofs, dark interiors usually habituated by wild animals, water trickling down through roof making surface slippery and difficult to walk, and so on. However, such derelict mines have been an ideal habitat for the rare species of insectivorous bats. There are several misconceptions about the bats, such as they bring bad omen, they are evil, or attract evil spirits; they suck blood and kill human beings, and so on. On the contrary, bats play an important role in ecology as prey and predator, for arthropod suppression, seed dispersal, pollination, material and nutrient distribution, and recycle, and have great advantages as well as disadvantages in economic terms. The economic benefits obtained from bats include biological pest control, plant pollination, seed dispersal, guano mining, bush meat and medicine, and education and research. Nevertheless, the misconceptions and negligence are high and the bats struggle for their survival. Species of bats like *Megaderma lyra lyra, Rousettus leschenaulti*, etc., are threatened because of increased urban development, deforestation and exploitation of caves. In this chapter, we present a general information about the bats and discuss the species of bats habituating abandoned underground Kandri manganese mine in Nagpur district of Maharashtra (India). As a measure to save the bats, it is important to educate people, especially locals, about the ecological importance of bats; in this regard, Geotourism is an effective tool.

M. Thakare
Sant Gadge Maharaj Mahavidyalaya, Hingna, Dist., Nagpur 441111, (MH), India

K. Randive (✉)
Department of Geology, RTM Nagpur University, Nagpur 440001, (MH), India
e-mail: randive101@yahoo.co.in

K. Randive et al. (eds.), *Innovations in Sustainable Mining*, Earth and Environmental Sciences Library, https://doi.org/10.1007/978-3-030-73796-2_14

Keywords Geological Tourism · Geoparks · Abandoned mines · Insectivorous bats · *Megaderma lyra lyra* · *Rousettus leschenaulti* · Survival of endangered bats

1 Introduction

Bats have long had an unfairly ghoulish reputation and often loathed as a weird monster. Popular culture links these nocturnal flying mammals to the dreaded vampires and Halloween decorations and considers them to be spooky. More recently, they have been accused of being a host for the origin of COVID-19 pandemic. The fact is, however, that bats help humans far more than they harm them. They rather help fight the spread of lethal diseases. Most bats are insect-hunters (insectivorous), capable of consuming enormous numbers of moths, flies, and mosquitoes. One bat can eat hundreds of mosquitoes in a single night. And they eat the types of mosquitoes that spread West Nile virus, zika, malaria, and other deadly pathogens. In certain places where bat colonies have collapsed, the number of mosquitoes has increased, increasing the disease prevalence [1].

Nevertheless, bats are the sources of economic benefits to human beings. Apart from controlling agricultural pests, they produce guano, which is an important natural fertilizer [2]. Chemical analysis of bat guano also gives historical evidence of climate change. Approximately one quarter of the world's 1,300 species of bats are reported on islands alone. Like other island mammals, island bat species pose a substantial threat of extinction due to loss of biodiversity or other threats. Nevertheless, derelict mines have always remained to be a favored habitat of the non-island continental bats. For example, the Quincy Mine, North America with bat-friendly gates is ideal site for bat tourism. Furthermore, Bracken Cave, near San Antonio considered as the largest bat colonies with 15–20 million bat residents [3]. The Congress Avenue Bridge in Texas Gunung Mulu National Park in Sarawak Malaysia, Carlsbad Cavern New Mexico, Kasanka National Park in Zambia; Citadel Berlin are some of the largest bat hotspots in the world [4, 5].

However, due to above reasons, the bats have become highly vulnerable species, and struggling for their survival. There have been several suggestions for the conservation of bats, however, change in the public perspective of the bats remain inevitable for their conservation. It is in this context that the geotourism in derelict mines gains importance. A locality-based guide to the species of bats can be very helpful for the nature enthusiasts to understand the ecology, biodiversity, and study of natural habitat of the animals of the area. More the number of people educated about the environment and ecology, more are the chances of preservation of endangered and rare species of animals like bats. It is in this view; this chapter presents an overview of the species of bats, their importance as a geotourism attraction, and a case study of the bats occurring in an abandoned manganese mine in Kandri area, Nagpur district of Maharashtra in central India.

2 Geotourism at a Glance

The term "Geotourism" is said to be coined by the National Geographic Society (NGS) and is a relatively new concept in the tourism industry that has emerged as a rapidly growing form of tourism which is being adopted worldwide [6]. Those who have a passion for knowledge, adventure, new discoveries, and all things real and amazing, National Geographic has long inspired curious and conscious people to explore the world. There are various ways in which 'Geotourism' is defined, most popular definitions are cited in Table 1.

All the definitions mentioned above indicate that the tourist's attraction towards the geologically and ecologically important natural landscapes, with or without being modified by the anthropogenic activities, should be included under the term Geotourism. The conservation of such sites is only possible, when they become the centers of economic activities. The mine sites are usually developed as the townships on which, the entire area delves upon for the commercial activities directly or indirectly related to mining. When such mines have lived their life and the mining activity no longer prevails in the area, the mines are abandoned and with time. Such mines become the favorable habitat of the animals such as bats, which love secluded, undisturbed, and remote locations. Usually, the abandoned mines are difficult to access due to non-maintenance of sites, waste water ponds that are often filled with unhealthy and contaminated waters, fragile waste dumps, anthropogenic wastes studded with sharp metallic tools such as nails, rods, machine parts, abandoned vehicles, plastic wastes such as containers, bottles, and, others. The abandoned mines are often inhabited by wildlife, bats, snakes, worms, and insects. There are many other possible dangers associated with the abandoned underground mines, namely, fragile roofs, dark interiors usually habituated by wild animals, water trickling down through roof making surface slippery and difficult to walk, and so on. All these factors make them less-attractive tourist destinations. However, there is a possibility of making them popular by targeting specific groups, such as knowledge mongers, the hardcore nature enthuziasts, the school children taken for study tour by their teachers and guides, the senior citizens, who are retired from active service but love knowledge-based tourism, and anyone else who has specific interests. For e.g., the zoologists, ornithologists, environmentalists, etc. will find the study of bats an interesting knowledge-based tourism attraction. Once the people start finding this new insight, these groups will further popularize such sites, and the number of visitors will gradually increase. This increase in number of tourists would propel the wheel of village economy, which was slowed or stopped due to discontinued mining activity. The tourism industry is one of the highest direct as well as indirect employment generation industry. Derelict mines are now considered as important tourist destinations [23]. There is a need to identify appropriate sites for geotourism, and create the base-line information so that the related agencies can develop these sites as tourist destinations.

Table 1 Popular definitions of Geotoursim

Definition	Authors
Geotourism is a knowledge-based tourism, an interdisciplinary integration of the tourism industry with conservation and interpretation of abiotic nature attributes, besides considering related cultural issues, within the geosites for the general public	Sadry [7]
Geotourism is fundamentally a geosite-based activity	Hose [8]
The geological and geomorphological tourism	Hose [9], Hose [10], Ollier [11]
A form of nature tourism focused on geological sites with emphasis on the "geological" element and geo-interpretation	Newsome and Dowling [12], Newsome et al. [13], Vasiljevic et al. [14], Gray [15]
Tourism that sustains or enhances the geographical character of a place-its environment, culture, aesthetics, heritage, and the well-being of its residents	National Geographic Society [16]
Geotourism, is a subcategory of tourism, which is considered as one of the new methods in providing tourism attraction	Servati and Qasemi [17]
Geotourism has a certain definition with geological tourism at its centre and deals with the investigation of related forms and consequences to earth, geomorphologic and geological phenomena	Newsome and Dowling [18]
Tourism in geological outlooks	Gates [19]
Geotourism deals with geology, geomorphology, natural outlooks and the forms of earth surface, layers with fossil, rocks and minerals with emphasis on the creating processes	Dowling and Newsome [20]
Geotourism is informed and responsible tourism in nature with the aim of looking at recognition of geological phenomena and processes and learning their formation and revolution	Amrikazemi [21]
Geotourism is not only a new part of tourism market, it is a principal guidance to help maintaining nature and sustainable development, which is compatible with the economic equilibrium, social condition and ecology and complements them	Rios-Reyes et al. [22]

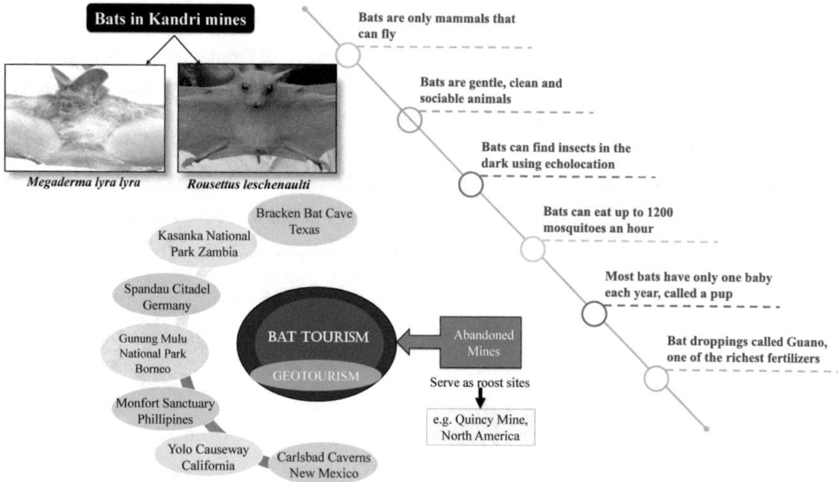

Fig. 1 Peculiar characteristics and popular sites of bats

3 Bats as a Potential Tourist Attraction

Bats are among the most popular members of the animal kingdom, if not for good reason, but have always been known to the human race. Most cave paintings have depiction of bats [24, 25]. Ancient as well as tribal cultures have several superstitions about the bats. Some of them believe in the supernatural powers of bats. Others believe that secretions by specific species of bats have great sexual powers, which fetches very high value for this material. That is why the locals risk their lives in winning this material by climbing very sharp and slippery cliffs [26]. In the vampire movies and Halloween decorations, bats find indispensable position. Bats are so popular that the popular comic character 'the Bateman' makes a superhero, who is loved by all children. In mainland China and some parts of Asia Pacific region, bats are included in the diet and make a popular food due to belief that it has a high medicinal value [27]. Due to this great deal of popularity, bats have always attracted people. Therefore, the bats make an excellent tourist attraction, some of the popular sites of bats are mentioned in Fig. 1.

4 General Information About Bats

Bats are the placental mammals occurring in the order Chiroptera. 'Bat' is the common name of the animal while Chiroptera is its scientific name. Phylum Mammalia and the order Chiroptera have been classically subdivided into two suborders, namely, Megachiroptera (Old World Fruit Bats) and Microchiroptera (laryngeally echolocating bats). However, with the application of molecular and phylogenetic methodologies, scientists have suggested a new subordinal division of Chiroptera, namely Yinpterochiroptera comprising megabat family Pteropodidae along with microbat families Rhinolophidae, Rhinopomatidae, Megadermatidae, and Yangochiroptera including rest of the microbat families [28, 29].

Pteropus (suborder Yinpterochiroptera) commonly known as Flying fox are the largest of all bat species with wings that span about five to six feet and weighing up to 2.2 lbs. Microbats, on the other hand, are highly specialized and are echolocating. The smallest of the bats is the Kittis hog-nosed bat, weighing about 2–2.6 g and measuring about 29–34 mm [30].

5 Importance of Bats

Bats is an important ecological creature, helpful in pest control, seed dispersal, pollination, arthropod suppression, nutrient distribution, and recycling, and have great economic benefits as well as drawbacks. However, in general, the potential benefits of bats are laid out below [31].

Integrated Pest Management: Insectivorous bats in temperate regions are highly competent predators which play a significant role in protecting economically valuable crops. Several research studies reported the presence of agricultural pest species in the bat diet [32]. Furthermore, bats are economically valuable for nontoxic pest-control operations in agriculture.

Plant Pollination: Plant-visiting bats bear a large amount of pollen on their face and fur, and offer important biological assistance by fostering regenerative efficiency and the emergence of new seedlings. Significant number of these plants are among the major species of biomass in their ecosystem [33].

Seed Dispersal: Owing to its unique species diversity, richness and a combination of canopy and understory feeding patterns, bats are considered as important seed dispersers. Organic product eating bats assume a critical part in timberland recovery. Tropical frugivorous bats also promote the regeneration of tropical timberland and help to preserve the diversity of plants by providing seeds from outside remote regions, whereas frugivorous neotropics play an important role in the early stages of forest progression. In various tropical backwoods, the dispersed seeds of palms and figs by bats are also common. Since they are likewise eaten by numerous winged animals and warm-blooded creatures, figs frequently serve as cornerstone species in tropical timberlands [34].

Prey for Vertebrates: The majority of the bats are originated before on perching or when they rise up out of perches albeit at times originated before scrounging or flying. Enormous convergences of bats at perch locales and the moderately unsurprising examples of their arising out of perches, give huge occasions to hunters to go after bats. Notwithstanding, techniques like low trustworthiness to perch destinations, determination of time, and examples of rising up out of perches and night-time movement are utilized to limit the danger of predation [35].

Bats as Predator: Predation of bats can affect herbivorous networks and circuitous implications for plant networks by collaborations between intervened thicknesses (utilization) and quality interceded (conduct) and for nature balance [36].

Bioindicator: Bats are incredible natural markers of territorial importance. They have immense potential as bioindicators for both aggravation and the prevalence of toxins because of a blend of their size, portability, life span, ordered steadiness, identifiable short and long-haul effects, population dynamics, and their circulation across the planet [37].

Hosts for Parasites: Various haematophagous ectoparasites reside on, for example, bat insects (Ischnopsyllidae), bat flies (Nycteribiidae), bat parasites (Spinturnicidae), and bugs (Cimicidae) on the skin surface and in the hide of bats. These commit ectoparasites are host-specific. The skin and hair morphology implies that vital parts influence the way of life of the parasite in terms of transition, feeding, growth and egg laying induced morphological differences with the coevolution of the two species [38].

Guano helps for Fertility and Nutrient Distribution: As bats regularly or sporadically perch in caverns, bat guano makes an important natural contribution to buckle biological processes that are inalienably deficient in essential quality. They provide basic natural info that characterizes arrays of diverse endemic cavern greenery. Cavern lizard and fish species as well as invertebrate networks, for example, are also strongly vulnerable to bat guano supplements [39].

Guano Mining: Guano from bats has long been mined from caves for use as fertilizer on organic crops due to its high levels of nutrient availability, such as nitrogen and phosphorus. It provides exquisite quality natural fertilizer in the world. Guano is mined in a sustainable manner in some areas, particularly in caves where bats typically migrate somewhere else for a part of their life each year. Few firms have also used the bacteria derived from bat guano to refine detergents and other items of considerable benefit to humans [40, 41].

Biological Pest Control: Bats have great role to play in the paste control, since the pests are part of regular diet of the insectivorous bats. Body organs of pastes were found in the guano of bats, which has provided important clue on the dietary mechanism of bats [42, 43].

Bat watching Tourism: Bats are indeed a pertinent part of the ecosystem and watching wildlife is altogether a recreational endeavour to witness their behaviour and enjoy their elegance [44].

Education and Research: Bats contributed a lot to the field of biomimetics, which is the science of modelling cutting-edge technologies based on natural forms. Anticlotting compounds in bat saliva are also being tested now as a possible anticoagulant for people at high risk of blood clots and strokes. Furthermore, the design of sonar for ships and ultrasound was partially inspired by the echolocation used by bats as a tracking device to locate and track their prey at night without falling on trees, buildings or other obstacles. [45, 46].

Disease Transmission and Contamination: Bats are reservoirs of several pathogens, whose spread may be related to physiological stress associated with habitat loss or alteration [47].

Bush meat and Medicine: Bat bush meat has the highest nutritional value (high mineral composition, protein and vitamin). An anticoagulant compound called salivary plasminogen activator (DSPA) present in the saliva of the common vampire bat is used to treat strokes. Doctors used bats to heal patients' ailments ranging from baldness to paralysis [48, 49].

6 World Scenario of Bats

Bats are common all over the world, except for the Arctic, Antarctic continents and a few small oceanic islands. Bats have a larger number of species after rodents (over 1200 species), which make up one-fifth of total mammalian population on earth. However, since there may be more species of bats in the world, these estimates are just relative indicators. Much of the bat population is located in the tropics, with South and Central America accounting for one-third of the total bat population in the world. The island country of Indonesia is a home for about 175 species of bats [30]. Chiropterans commonly known as bats are the only true flying mammals, comprising together 1116 species in 202 genera under 18 families globally and constitute about a quarter of the entire mammal species [50]. Table 2 gives the details of number of species of bats found around the world.

7 Bats in India

India is one among the 25 mega-biodiversity hotspots that harbor the richest and most highly endangered eco-regions of the world. Chiropterans (including Mega- and Microchiroptera) are significant because they are widely distributed in almost every state of India. They play important role in all types of ecosystem [51]. India is home to about a hundred species of bats, including 12 fruit bats, such as the fulvous fruit bat *Rousettus leschenaulti*, Nicobar flying fox *P. faunulus*, Blyth's flying fox *Pteropus melanotus*, island flying fox *P.* hypomelanus, Indian flying fox *P. giganteus,* short-nosed fruit bat *Cynopterus sphinx*, lesser dog-faced fruit bat C. *brachyotis*, Ratanaworabhan's fruit bat *Megaerops niphanae*, Salim Ali's fruit bat *Latidens salimalii*, Blanford's fruit bat *Sphaerias blanfordi*, dawn bat *Eonycteris spelaea*, and hill long-tongued fruit bat *Macroglossus sobrinus* (Table 2). Nevertheless, the data on the conservation status, population density, and ecology of many of these species is

Table 2 Population of Bat Species in world and their abundance in India

Names of families of bats	No. of species in world	No. of species in india
Pteropodidae	190	14
Rhinolophidae	77	18
Hipposideridae	81	16
Megadermatidae	5	2
Rhinopomatidae	4	2
Craseonycteridae	1	–
Emballonuridae	51	6
Nycteridae	16	–
Myzopodidae	1	–
Mystacinidae	2	–
Phyllostomidae	160	–
Mormoopidae	10	–
Noctilionidae	2	–
Furipteridae	2	–
Thyropteridae	3	–
Natalidae	8	–
Molossidae	100	4
Vespertilionidae	407	62
Miniopteridae	0	3

limited due to lack of field studies. Numbers of bat species reported in India are: 9 Families, 39 Genera, 117 species and 100 subspecies [52].

8 Bat Species in Abandoned Underground Manganese Mine at Kandri, Nagpur District, Maharashtra

Maharashtra state is the third largest by area in India. Eight families, 23 genera, and 41 species characterize the bat fauna of Maharashtra, most of which are insectivorous microchiropterans. Eleven of the 41 species of bats are distributed across the state, while 21 species have few colonies with limited distribution. Five species endemic to South Asia occur in the state. Chiropteran in Maharashtra by and large favor forest habitat; many caves in these areas are ideal as roosting sites. Nagpur region has remarkably high forest cover (34.84%) but poorly known bat fauna; therefore, more surveys are recommended to determine bats species richness in that area [53]. Kandri Manganese Mine owned by the Manganese Ore India Limited (MOIL), is situated at about 42 km from Nagpur, a major city in Central India. Present study was carried out to investigate the bat species dwelling in the derelict Kandri underground mine

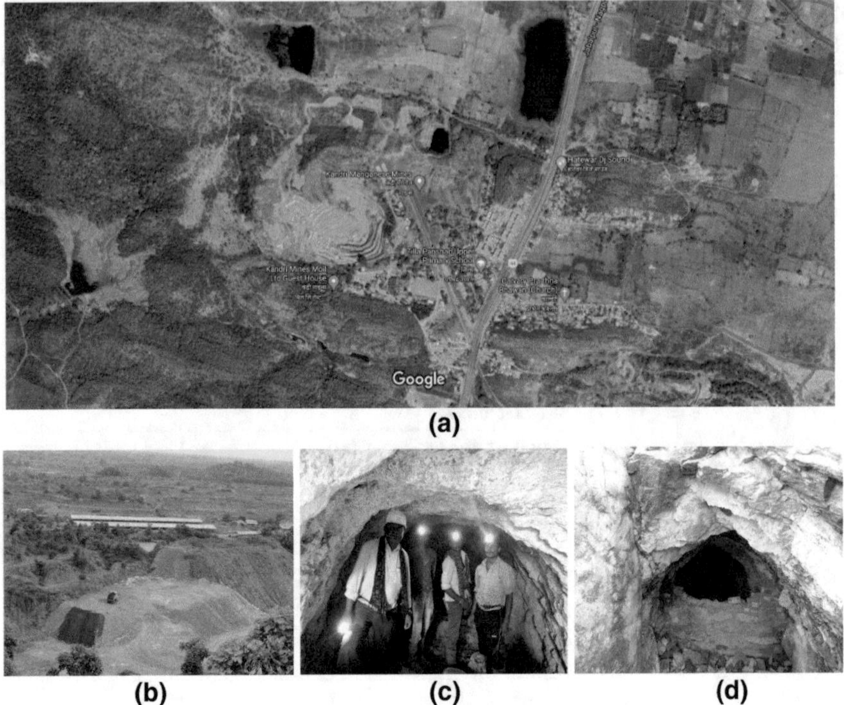

Fig. 2 **a** Google™ map showing location and present-day level of excavation at Kandri Mines, **b** view of mined benches and mine dumps, **c** inside view of underground mine, **d** Abandoned tunnel of underground Kandri mine

(Fig. 2). Two species, namely *Megaderma lyra lyra* and *Rousettus leschenaulti* are found here. Detail description of these species along with their breeding habits, and their protection status are given below.

8.1 Megaderma Lyra Lyra *(Greater False Vampire Bat)*

The *Megaderma lyra lyra* is found inhabiting old abandoned caves, mines, old houses, and cow shades. *Megaderma lyra lyra* are widely distributed in colonies, and these species are very rare and stay far from each other. This species is, however, distributed in many states of India. In Maharashtra, it has been located at Fort of Dualatabad and Kandri Mines, Nagpur district. They are also found at Mandu near Indore in Madhya Pradesh. The species can be identified by a strong penetrating odour of ammonia with deposition of guano on the floor of inhabited sites. *Megaderma* is a large insectivorous bat and selects its roosting place away from strong light with high humidity and high temperature and fewer disturbances. The population of colony may vary from 50–80 individuals up to those numbering hundreds. No segregation of the specimen based on sex, age or season was noticed. This species can share their roosts with other bat species [54].

This species is 6.4–9.6 cm (2.5–3.7 in) in length and weighs 38 to 55 g. The body size is relatively big. The average forearm length is about 6.6–7.0 cm (2.6–2.7 in). Dorsal fur of this bat is mouse-grey. Ventral fur paler. Juveniles are darker. The ears are wide, joint medially for a third of their length, and possess a bifid tragus. The noseleaf is erect and 9 mm in length. It has large ears and no tail (Fig. 3a).

M. Lyra uses a variety of tactics to hunt. Approximately 85% of the prey is trapped during short flight searches in which it flies about half a meter above the ground level. It also employs a sit-and-wait technique to hunt the prey perching about two meters off the ground. The diet of this species consists predominantly of insects and small vertebrates, such as amphibians (frogs and toads), reptiles (lizards), fishes, birds (white-eyes, sunbirds, sparrow, dusky crag martin) and mammals (bats, rats, mice, gerbils). It uses echolocation. It can hunt with both vision and echolocation, thus being able to catch prey in total darkness.

Breeding Habit: The breeding habits of *Megaderma lyra lyra* have been described in detail by Gopalkrishna [55–57]. This species is the seasonal breeder. It has restricted breeding season and breed once in a year. The *Megaderma lyra lyra* is a seasonally monoestrous species. The males and the females have almost a synchronous sexual season, with the males being slightly ahead of the females in each year. The species has annual reproductive cycle extending from August to November. Microscopic observation of the testes shows that spermatogenesis commences in September, spermiation occurs in October and vigorous spermatogenesis occur in November while the regression changes start in the testes from December. From January to August, bats are in quiescence phase where no spermatogenesis occurs in testes. The reproductive cycle of adult male *Megaderma lyra lyra* are as follows.

(1) Sexually quiescence or inactive (July to September). Testes do not show spermatogenesis. The testes size is reduced and no sperms in epididymis.

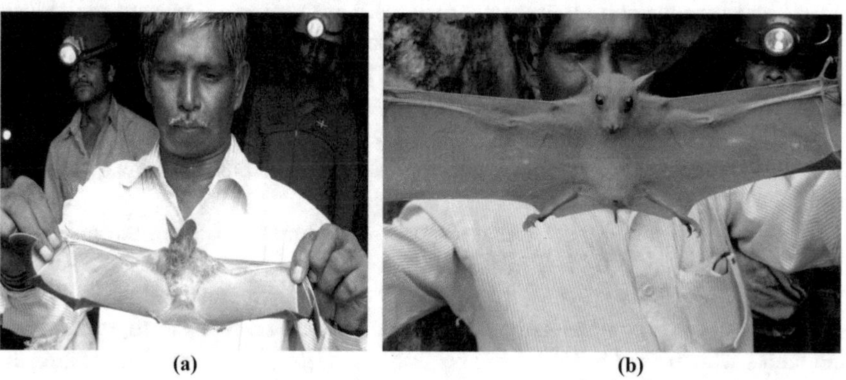

(a) (b)

Fig. 3 Bats species found in Kandri Mines **a** *Megaderma lyra lyra and* **b** *Rousettus leschanaultii*

(2) Pre-breeding or recrudescence (Late September to October). Testes show initiation of spermatogenesis and spermiation but no sperms in epididymis.
(3) Sexually active or breeding (mid-November). Testes show vigorous spermatogenesis and sperms are present in the epididymis.

The female bat is in inactive period up to September. From the last week of September, the activities in the ovary start which show the development of follicles. The follicles grow to the vesicular level up to October and the follicles attain the stage of Graafian follicle in the left ovary only. As the left side has the dominancy over the right. The right ovary at this stage shows the atretic follicles. Copulation taken place in the 1st week of November; which then results in fertilization leading to pregnancy. A single young one is born about the third week of April. The period of gestation is 150–160 days. The development of embryo takes place progressively during further months up to April, delivery takes place in the first week of April, and lactation continues up to May, during which the young ones is attached to the body of mother. The development of embryo mostly takes place in the left uterine cornu, but only in few exceptional cases it can happens in right side too. Thus, in *Megaderma lyra lyra* there is totally the physiological dominance of left side over the right side. The period of lactation extends up to latter half of June when the young ones leave their mothers. The young ones grow rapidly in size, but do not reach sexual maturity within the year of their birth [55, 56].

Status and Protection: Greater false vampire bats are at LR/lc, assessed by the Red List of Threatened Species [60]. They are widely distributed in some areas of India.

8.2 Rousettus Leschenaulti (*Fulvous Fruit Bat*)

This species has a better environmental tolerance. In a variety of habitats, from tropical forests to urban areas, this bat species is evident. It is found in caves, old derelict houses and tunnels, and other such structures. Up to several thousand individuals may be part of a colony of this species. In the thick canopy of a large-leafed tree or palm, solitary males would sometimes be found. This species may share their roosts with other bat species. *Rousettus leschenaulti* is the Indian megachiropteran tropical fruit-eating bat which lives practically in complete darkness or in very low light intensities, ideally available in abandoned mines. It feeds on flowers, fruits and nectar.

The head and body length measures 10.5–11.5 cm (4.1–4.5 in), the tail length measures 1.2–1.9 cm (0.47–0.74 in) and the forearm length is about 7.1–8.2 cm (2.7–3.2 in).

The dorsal fur is completely brown, covering the crown of the head, back, throat and flanks whereas the ventral fur is grey or brown-grey, more greyish towards the centre. It has long pale hairs under the chin, an elongated muzzle and large dark eyes (Fig. 3b).

Table 3 Conservation Status of Bats as of 2012 and 2020 according to IUCN red list (1314) Species in total 233

	2012	2020
Critically endangered	–	1.6%
Endangered	–	6.3%
Vulnerable	–	8.3%
Threatened	15%	–
Near threatened	7%	6.7%
Least concern	–	58.0%
Data deficient	18%	18.4%
Extinct	0%	0.7%

Breeding Habits: The breeding habits and the reproductive cycle of adult male *Rousettus leschenaulti* are as follows.

(1) Sexually quiescence or inactive (June to August and December to February). Testes do not show spermatogenesis. The testes size is reduced and no sperms in epididymis [59].
(2) Prebreeding or recrudescence (September and March). Testes show initiation of spermatogenesis and spermiation but no sperms in epididymis.
(3) Sexually active or breeding (October–November and April- May).

Life history: In some areas of India this species can undergo two pregnancies in a year Young are actually born at 12 g, and suckling can last 35–40 days. Females may attain sexual maturity at 5 months, males at 15 months [58].

Emerges about 45 min after sunset in Madhya Pradesh, India. The diet mainly consists of fruits. Dietary overlap with *Cynopterus sphinx* was relatively high in the rainy season (June–October), while the species left the area in the dry season (November–December) when food was scarcer. It is also reported that they can occasionally feed on fish in India. This species emits clicks by running its tongue over the roof of its mouth. It produces echolocation clicks range from 60–12 kHz with most energy at around 18 kHz.

Status and protection: This species is regarded as pest mammal in the area of fruit farms, because they eat fruit. Fulvous fruit bats are at LR/lc, assessed by the Red List of Threatened Species [60]. They are widely distributed and common in some areas of southern Asia. Bats are important for seed dispersal. Possible threats may be habitat loss through development, dams, and deforestation, is also hunted for medicinal purposes [60].

9 Conservation Status and Remedies

Current status of survival of bats is far from good; Table 3 shows the IUCN statistics in 2012 and 2020, which also shows increased number of critically-endangered, endangered, and vulnerable species. Bat populations in India are precariously balanced on the basis of their survival. Attitude towards bats, myths about them, reckless hunting, disturbance of their natural habitat, and lack of legal framework for protection of bats are major impediments in their survival. Nevertheless, following steps could be very useful for their conservation. First and foremost is to educate the people about bats and burst the myths and misconceptions about them. Importance of bats should be made known to the people at large. Secondly, planning of proper conservation projects, complete ban on bat hunting, and creating roosting sites for bats can make lot of difference. Very little is known about the population status of various species of bats, however new data has been continuously inundated. Collection and propagation of these data will be useful for conservation purpose. Thirdly, a critical role has to be paid by the local legislators and parliamentarians, who can support, facilitate, and monetize the projects for conservation. Time is also ripe for revival of the legislation [60] to include protection to all species of frugivorous bats, and insectivorous bats.

10 Suggestions for Development of Kandri Mine as a Geotourism Attraction

Kandri mine is located very near to Nagpur city, which is a major city in Central India having an international airport, central railways main junction, and good network of national and state highways. Kandri mine is also very near to several archaeological sites at Mansar, Nagardhan, and famous temple of Lord Rama at Ramtek, and Khindsi lake which are already well-established tourist destinations. This location advantage can be very useful for developing Kandri mine as a tourist destination for bat watching. Another advantage is that the area has good road network, approach to the site is direct, however needs to be developed further for smooth access of the vehicles. Near the entrance of mine, there should be an area having tourist information centre with basic amenities such as eateries and lavatories. A tourist information centre can have local guides, small information booklets, and hoardings giving information about the bats. One such hoarding is given in Fig. 4.

11 Summary

It is possible to develop abandoned Kandri underground mine for geotourism, and there are many advantages associated with it, which are listed below.

(a) Availability of infrastructure for easy access of the site.

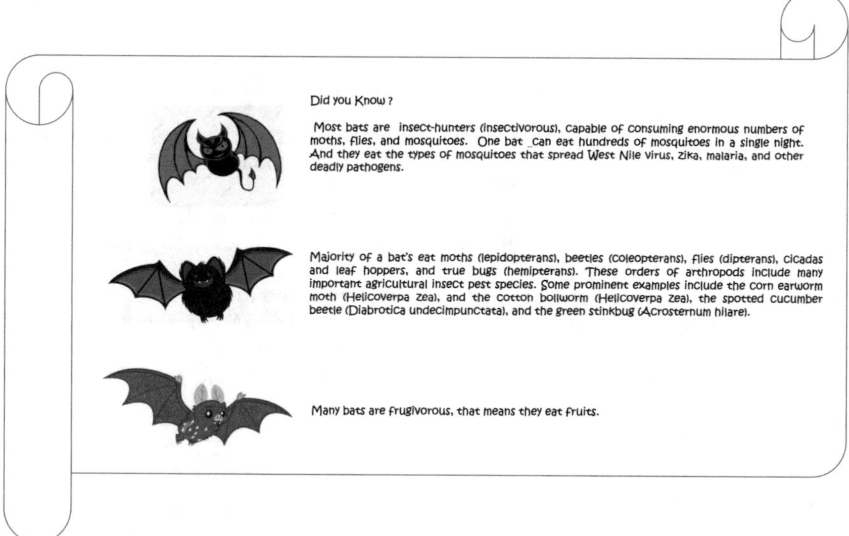

Fig. 4 Proposed hoarding that can be displayed at the Kandri geotourism site

(b) Availability of other popular tourist destinations is a close proximity of proposed geotourism site.
(c) Capacity of the subject (study of bats) to attract tourists, nature enthusiasts, school and college children, animal lovers, and other groups of tourists.
(d) Availability of resource persons for guiding interested students for pursuing further study on bats.

Geotourism provides a great opportunity for conservation of the fast-dwindling bats in India. This activity will provide a replicable model in other parts of the world.

References

1. Brooke M, Kathleen K (2020) Ecological and economic importance of bats in integrated pest management. Cooperative Extension Fact Sheet FS1270. https://njaes.rutgers.edu/fs1270/
2. Randive K, Raut T, Jawadand S (2021) An overview of the global fertilizer trends and India's position in 2020. Miner Econ. https://doi.org/10.1007/s13563-020-00246-z
3. Bat Tourism: The surprising power of bat-watching holiday makers bat conservation international is a 501(c)(3) organization
4. Pennisi L, Holland SM, Stein TV (2004) Achieving bat conservation through tourism. J Ecotour 3(3):195–207. https://doi.org/10.1080/14664200508668432
5. Bagstad K, Wiederholt R (2013) Tourism values for mexican free-tailed bat viewing. Human Dimen Wildlife 18(4). https://doi.org/10.1080/10871209.2013.789573
6. Turner S (2005) Promoting UNSECO global geoparks for sustainable development in the Australian-Pacific region. Alcher Austral J Paleontol 30(1):351–365

7. Sadry BN (2014) Fundamentals of geotourism: with special emphasis on Iran, 3rd Edn. Samt Organization publishers Tehran, p 220. https://physio-geo.revues.org/4873?file=1(geotourism
8. Hose TA (2011) The English origins of geotourism (as a vehicle for geoconservation) and their relevance to current studies. Acta Geogr Slovenica 51(3):343–359
9. Hose TA (1995) Selling the story of Britain's stone. Environ Interpret 10(2):16–17
10. Hose TA (2000) European geotourism–geological interpretation and conservation promotion for tourists. In: Barettino D, Wimbledon WAP (eds) Geological Heritage: Its Conservation and Management. ITGE, Madrid, pp 127–146
11. Ollier C (2012) Problems of geotourism and geodiversity. Quaestiones Geogr 31(3):57–61
12. Newsome D, Dowling RK (2010) Geotourism: the tourism of geology and landscape. Goodfellow Publishers, Oxford
13. Newsome D, Dowling RK, Leung YF (2012) The nature and management of geotourism: a case study of two established iconic geotourism destinations. Tourism Manag Perspect 2–3:19–27
14. Vasiljevic DA, Markovic SB, Hose TA et al (2011) The introduction to geoconservation of loess-palaeosol sequences in the Vojvodina region: signifcant geoheritage of Serbia. Quatern Int 240(1–2):108–116
15. Gray M (2013) Geodiversity. Chichester, Wiley Blackwell, p 495
16. National Geographic Society (2015) Geotourism Principles. https://www.nationalgeographic.com/travel/sustainable-destinations-photos/
17. Servati MR, Qasemi A (2008) Geotourism strategies in fars. J Geogr Space 2:6
18. Newsome D, Dowling RK (2006) The Scope and Nature of Geotourism. In: Newsome D, Dowling RK (eds) Geotourism, Chapter One. Elsevier, Oxford, pp 3–25
19. Gates AE (2006) Geotourism: A Perspective from the USA. In: Dowling RK, Jiménez-Sánchez M (eds) Quantitative indexes based on geomorphologic features: a tool for evaluating human impact on natural and cultural heritage in caves. J Cultural Heritage 12:270–278
20. Dowling RK, Newsome D (2006) Geotourism's issues and challenges. Geotourism, Chapter Thirteen. Elsevier, Oxford, pp 242–254
21. Amrikazemi A (2009) Atlas of geopark & geotourism resources of Iran. Geological Survey of Iran Publication, Tehran, pp 22–23
22. Ríos-Reyes CA, Manco-Jaraba DC, Castellanos-Alarcón OM (2018) Geotourism in caves of Colombia as a novel strategy for the protection of natural and cultural heritage associated to underground ecosystems. Biodiversity Int J 2(5):464–474. https://doi.org/10.15406/bij.2018.02.00101
23. Mendes I (2018) Derelict mining sites: environmental menaces and social cemeteries, or opportunities for local sustainable development? an essay 1. In: Utopia, Anarquia e Sociedade, Almedina Project: Contribution of Social Responsibility to Sustainable Development, pp 325–342
24. National Geographic society (2019) Ancient Australian rock art depicts unknown bats. https://blog.nationalgeographic.org/2008/12/09/ancient-australian-rock-art-depicts-unknown-bats/
25. Rincon P (2008) Rock painting reveals unknown bat. https://news.bbc.co.uk/2/hi/science/nature/7765136.stm
26. Bittel J (2013) Bats are the most fascinating, bizarre, Generous, Sexy Beasts. https://slate.com/technology/2013/10/bizarre-bat-behavior-oral-sex-pollinating-tequila-sharing-meals-drinking-blood-males-lactating.html
27. Mildenstein T, Tanshi I, Racey PA (2016). Exploitation of bats for bushmeat and medicine. bats in the anthropocene: conservation of bats in a changing world. Springer, p 327. https://doi.org/10.1007/978-3-319-25220-9
28. Springer MS, Teeling C, Madsen O, Stanhope MJ, de Jong WW (2001) Integrated fossils and molecular data reconstruct bat echolocation. Proc Natl Acad Sci United States Am 98:6241–6246
29. Teeling EC, Madsen O, Stanhope MJ, de Jong WW, Van Den Bussche R, Springer MS (2002) Microbat paraphyly and the convergent evolution of key innovation in Old World rhinolophoid microbats. Proc Natl Acad Sci United States Am 99:1431–1436
30. Sundra Chelsea Atitwa on January 15 2018 in World Facts

31. Mohammad K, Mundanthra B (2013) Ecological and economical importance of bats (Order Chiroptera) Hindawi Publishing Corporation. ISRN Biodiver, vol 2013, Article ID 187415
32. Kunz TH, Torrez EZ, Bauer D, Lobova T, TH Fleming (2011) Ecosystem services provided by bats Ann NY Acad Sci 12231–12238. https://doi.org/10.1111/j.1749-6632.2011.06004.x
33. Arizaga S, Ezcurra E, Peters E, Arellano FR, Vega E (2000) Pollination ecology of agave macroacantha (Agavaceae) in a Mexican Tropical Desert: the role of pollinators. Am J Botany 87(7):1011–1017
34. Muscarella R, FlemingTH (2007) The role of frugivorous bats in tropical forest succession. Biol Rev 82(4):573–590
35. Speakman JR (1991) The impact of predation by birds on bat populations in the British Isles. Mammal Rev 21(3):123–142
36. Schmitz OJ, Suttle KB (2001) Effects of top predator species on direct and indirect interactions in a food web. Ecology 82(7):2072–2081
37. Jones G, Jacobs D, Kunz TH, Wilig MR, Racey PA (2009) Carpe Noctem: the importance of bats as bioindicators. Endang Species Res 8:3–115
38. Dittmar K, Porter ML, Murray S, Whiting MF (2006) Molecular phylogenetic analysis of nycteribiid and streblid bat flies (Diptera: Brachycera, Calyptratae): implications for host associations and phylogeographic origins. Molecul Phyl Evol 38(1):155–170
39. Fenolio DB, Graening GO, Collier BA, Stout JF (2006) Coprophagy in a cave-adapted salamander; the importance of bat guano examined through nutritional and stable isotope analyses. Proc R Soc B 273(1585):439–443
40. Hutchinson GE (1950) Survey of existing knowledge of bio- geochemistry: the biogeochemistry of vertebrate excretion. Bulletin Am Museum Natural History 96:1–554
41. Hougner C, Colding J, derqvist TS (2006) Economic valuation of a seed dispersal service in the Stockholm National Urban Park, Sweden. Ecol Econ 59(3):364–374
42. Coutts RA, Fenton MB, Glen E (1973) Food intake by captive *Myotis lucifugus* and *Eptesicus fuscu*s (Chiroptera: Vespertilion- idae). J Mammal 54:985–990
43. Kurta A, Whitaker JO (1998) Diet of the endangered Indiana bat (*Myotis sodalis*) on the northern edge of its range. Am Midland Natural 140(2):280–286
44. Tapper R (2006) Wildlife watching and tourism: a study on the benefits and risks of a fast-growing tourism activity and its impacts on species, UNEP/CMS Secretariat, Bonn, Germany
45. Bunget G, Seelecke S (2010) BATMAV: a 2-DOF bio-inspired flapping flight platform in The International Society for Optics and Photonics, vol 7643 of Proceedings of SPIE, pp 1–11
46. Müller R, Kuc R (2007) Biosonar-inspired technology: goals, challenges and insights. Bioinspir Biomimet 2:146–161
47. Fenton MB, Davison D, Kunz TH, McCracke GF, Racey PA, Tuttle MD (2006) Linking bats to emerging diseases.Science 311(5764):1098–1099
48. Jenkins RKB, Racey PA (2008) Bats as bushmeat in Madagasca. Madagascar Wildlife Conserv 3:22–30
49. Mickleburgh S, Waylen K, Racey P (2009) Bats as bushmeat: a global review. Oryx 43(2):217–234
50. Wilson DE, Reeder DM (2011) Class mammalia linnaeus, 1758. In: Zhang ZQ (ed) Animal biodiversity: an outline of higher-level classification and survey of taxonomic richness. Zootaxa 3148:56–60
51. Rahman A, Choudhury P (2008) Department of ecology and environmental science, Assam
52. Talmale S, Pradhan MS (2009) A checklist of valid indian bat species (Chiroptera: Mammalia) Zoological Survey of India
53. Korad VS (2014) Studies on diversity, distribution, and conservation of the bat fauna in Maharashtra State, India Aprobanica, And Issn 1800–427x0. June, 2014, 06(01):32–45
54. Thakare MU (2014) Thesis. Rashtrasant Tukdoji Maharaj Nagpur University, Nagpur (RTMNU)
55. Gopalakrishna A, Khaparde MS (1978) Early development, implantation and amniogenesis in the Indian vampire bat, Megaderma lyra lyra (Geoffroy). Proc Animal Sci 87(6):91–104. https://doi.org/10.1007/BF03179267

56. Gopalakrishna A, Khaparde MS (1978) Development of the foetal membranes and placentation in the Indian false vampire bat, Megaderma lyra lyra (Geoffroy). Proc Animal Sci 87(9):179–194. https://doi.org/10.1007/BF03179005
57. Gopalkrishna A, Choudhari PN (1977) Breeding habits and associated phenomenon in some Indian bat. Part 1–*Rousettus leschenaulti* (Desmarest) Megachiroptera. J Bombay Nat Hist Soc 74:1–16
58. Karim KB, Wimsatt WA (1979) Electron microscopic observations on the yolk sac of the Indian fruit bat, Rousettus leschnaulti (Desmarest) (Pteropidae). Anatom Record 195(3):493–510. https://doi.org/10.1002/ar.1091950309
59. Gopalakrishna A, Karim KB (1974) Secretory blebs from the seminal vesicles of the indian fruit bat, Rousettus leschenaulti (Desmarest). Current Sci 43(12):383–412
60. Hussain Z (2017) Environmental legislation. The Statesman Archived from the original on 30 November 2020

Printed in the United States
by Baker & Taylor Publisher Services